Violence in Extreme Conditions

Eric-Hans Kramer · Tine Molendijk
Editors

Violence in Extreme Conditions

Ethical Challenges in Military Practice

Springer

Editors
Eric-Hans Kramer
Faculty of Military Sciences
Netherlands Defence Academy
Breda, The Netherlands

Tine Molendijk
Faculty of Military Sciences
Netherlands Defence Academy
Breda, The Netherlands

ISBN 978-3-031-16118-6 ISBN 978-3-031-16119-3 (eBook)
https://doi.org/10.1007/978-3-031-16119-3

This Springer imprint is published by the registered company Springer Nature Switzerland AG
The registered company address is: Gewerbestrasse 11, 6330 Cham, Switzerland

Acknowledgements

This book was written for Desiree Verweij's farewell as Professor of Military Ethics at the Netherlands Defence Academy. Several colleagues, former colleagues, former Ph.D. students and fellow professors address the theme of "violence in extreme conditions" from their specific expertise and theoretical perspectives. This theme has been central to Desiree's approach to military ethics.

We would like to thank several people for their contribution to the creation of this book. In addition to, of course, the authors, whom we thank for investing time in elaborating their specific perspectives on this theme, we thank Johannes Glaeser and Yogesh Padmanaban of Springer for their constructive and patient approach and their trust in us as editors. In addition, we thank Mark Vollmer, Arthur Eveleens and Aurélia Kanounji of the Translation Department of the Defence Language Centre for their support in proofreading the chapters. Finally, the Netherlands Defence Academy's generous funding of the open access costs for this book is sincerely appreciated.

Reflection on violence in extreme conditions is essential for the military organization. At the Netherlands Defence Academy, moreover, our future officers receive their training, which makes a reflection on the topic particularly relevant for our academy. Desiree has put this theme on the map within the academy, but we firmly intend to continue working on it. The enthusiasm of the authors to contribute to this book underscores that it is a fruitful theme that can be approached integratively from many angles. Therefore, this book not only looks back at Desiree's merits, but we cast a glance at the future in which we hope to continue and advance our research and education on this violence in extreme conditions.

Breda, 2022

Eric-Hans Kramer
Tine Molendijk

Acknowledgements

Contents

Editors and Contributors

About the Editors

Prof. Eric-Hans Kramer is Professor in Management and Organization Studies at the Netherlands Defence Academy. Before that, he was associate professor of Human Factors and Systems Safety at the same organization. He is involved in several research projects into organizational issues the military organization is being confronted with and with the development of research and educational programs into the psychosocial dynamics in military organizations.

Dr. Tine Molendijk is an interdisciplinary-minded cultural anthropologist specialized in the topics of violence, military culture, and mental health, in particular post-traumatic stress disorder (PTSD) and moral injury. Her Ph.D. research focused on the role of political practices and public perceptions in moral injury in Bosnia and Afghanistan veterans. Currently, she works as Assistant Professor at the Faculty of Military Sciences at the Netherlands Defence Academy (NLDA) and as a Research Fellow affiliated to the Radboud University. Among her publications is the award-winning book *Moral Injury and Soldiers in Conflict: Political Practices and Public Perceptions* (Routledge, 2021).

Contributors

Braun Christian Nikolaus Radboud University, Nijmegen, The Netherlands

Eikenaar Teun Radboud University, Nijmegen, The Netherlands

de Graaff Miriam Carla Royal Netherlands Armed Forces Joint Centre of Expertise Communication & Engagement, Apeldoorn, The Netherlands

Grassiani Erella Department of Anthropology, University of Amsterdam, Amsterdam, The Netherlands

Jansen Marenne Head Strategic Unit, Ministry of Defense Netherlands, Hague, The Netherlands

Karssing Edgar Nyenrode Business Universiteit, Breukelen, The Netherlands

Kramer Eric-Hans Department of Military Management Studies, Faculty of Military Sciences, Netherlands Defense Academy, Breda, The Netherlands

Molendijk Tine Department of Military Management Studies, Faculty of Military Sciences, Netherlands Defense Academy, Breda, The Netherlands

Moorkamp Matthijs Institute for Management Research, Nijmegen School of Management, Radboud University, Nijmegen, The Netherlands

Olsthoorn Peter Netherlands Defence Academy, Breda, The Netherlands

Peperkamp Lonneke Faculty of Military Sciences, Netherlands Defence Academy, Breda, The Netherlands

van Tongeren Paul Department of Philosophy, Radboud University, Nijmegen, The Netherlands

Verweij Desiree Faculty of Military Sciences, Netherlands Defence Academy, Breda, The Netherlands;
CICAM, Radboud University, Nijmegen, The Netherlands

Visser Max Institute for Management Research, Nijmegen School of Management, Radboud University, Nijmegen, The Netherlands

Zalm Claire Ministry of Foreign Affairs, Department of Security, Crisis Coordination & Integrity, The Hague, The Netherlands

Introduction

Eric-Hans Kramer and Tine Molendijk

'You know what the issue is with the military academy? Nobody ever talks about violence'. This observation, made by Désirée Verweij at the beginning of this millennium, initiated the development of a particular approach to military ethics. This approach eventually found its way into various educational programmes and research projects with considerable success. What is so special about this observation? Why should a seemingly offhand remark in hindsight be brought forward as a defining moment in the development of the chair in military ethics at the Netherlands Defence Academy, which was held by Désirée Verweij for more than a decade? Is it even true? It might seem unlikely that the very topic that is ignored at the military academy is violence. After all, the potential confrontation with violence in extreme conditions is a distinctive feature of military practice. In this introduction, we reflect on Désirée's observation and the questions it triggers in order to show why it can bring together a broad array of authors with backgrounds in different fields, such as philosophy, ethics, anthropology, psychology, organization science and law, around the topic of 'the confrontation with violence'. Interdisciplinary explorations of this theme have been at the core of the work of Désirée, who held the chair in military ethics at the Netherlands Defence Academy from 2008 to 2021. The contributions in this book celebrate her achievements and seek to keep important themes in the spotlight.

So, is violence really never talked about at the military academy? To start with, we should emphasize that Désirée's observation was aimed at the academic programmes, not necessarily at the military training that is also part of officer education. Furthermore, violence is of course talked about in various curricula. Military practice is a reference point for bringing coherence to the different academic programmes and research projects. Particularly the interplay between academic theory and military

E.-H. Kramer (✉) · T. Molendijk
Faculty of Military Sciences, Netherlands Defence Academy, Breda, The Netherlands
e-mail: fj.kramer@mindef.nl

T. Molendijk
e-mail: t.molendijk@mindef.nl

E.-H. Kramer and T. Molendijk (eds.), *Violence in Extreme Conditions*,
https://doi.org/10.1007/978-3-031-16119-3_1

practice establishes the relevance of both education and research at the Netherlands Defence Academy. Yet, Désirée's point was that violence is primarily talked about in a specific way. As one of the organizations that executes the state's monopoly of force, the military organization is expected to use violence in a technical and instrumental way. It is the instrumental application of force that defines the legitimacy of military practice. Topics such as 'air power', 'hybrid challenges', 'doctrinal developments', 'behavioural engineering', 'human enhancement', 'legitimacy', 'technological innovation', 'collateral damage' and the like are certainly part of everyday conversation. However, Désirée's point was that a one-sided emphasis on the instrumental application of force and the technical rationality that accompanies it is limiting and might turn attention away from the brutal reality that is also part of the military profession.

Violence is a constitutive element of this brutal reality and plays a critical role in various ways, and in various guises, in military practice. As an organization that operates in extreme conditions, the military organization may be confronted with the destructive behaviour of individuals, organizations and societies. The military may be confronted with abuses of power and aggression, as well as with the consequences of such behaviour, such as human suffering and anxiety. Moreover, the military organization may trigger violent and destructive behaviour within its own ranks, and may create the conditions in which trauma in individual service members can develop. This explains why violence forms a point of orientation for ethical reflection on military practice, and it is the reason why Désirée's observation at the beginning of this millennium was significant. It also shows why such a perspective might be generative of a multitude of different disciplinary perspectives that can add specific themes and arguments. Furthermore, it shows why keeping an open mind to different perspectives in order to bridge and connect them is important to Désirée's approach. This volume therefore specifically aims to bring together a broad array of authors around this theme.

Désirée Verweij's Philosophical Position in Relation to Ethics and Violence

How does Désirée's own position as a philosopher connect to this theme of military practice and violence? As an important disclaimer for what follows, it should be emphasized that this is our own reconstruction, which we have developed by working together with Désirée and discussing these issues over the past years. There is no doubt that she would bring up several counterpoints to what we are laying out here, but we have deliberately chosen not to consult her for this reconstruction. After all, who would like to comment on a text that is written in celebration of one's own achievements? Any misrepresentations are therefore our responsibility.

Strongly influenced by Sigmund Freud, Friedrich Nietzsche and Hannah Arendt, Désirée is inclined to look for ambiguity; for ways in which people are in internal conflict, specifically because military ethics, which focuses on ethical reflection in an organization that is entrusted with executing the state's monopoly of force, is inevitably confronted with people's ambiguous and paradoxical relationship with violence (Verweij, 2007). This is directly related to Nietzsche's reflections on the topic (Verweij, 1999, translation EHK, TM): 'Because according to Nietzsche it is not at all clear what we are, but we are certainly no unity. We are a collection of "wills" says Nietzsche, and these "wills" are continuously in conflict. The "I" we are talking about is continuously changing'. To Désirée, the ambiguous nature of human beings is an existential foundation that informs her perspective.

In what specific way have Freud, Nietzsche and Arendt been inspirations to Désirée's philosophical position in relation to ethics and violence? Below, we discuss different themes that relate to these inspirations and that relate to contemporary discussions of military practice and violence. We discuss the way that they informed Désirée's position and the way that they are currently relevant to educational and research programmes at the Netherlands Defence Academy. We do not wish to claim that Désirée is a Freudian or a Nietzschean. Too much valid critique is available on, for example, Freud, and the different debates have become too elaborate to attribute Désirée's outlook to a single reference point. Typical for Désirée's style is that, without denying and dismissing critique, one may appreciate valuable insights that can be found everywhere, but for her particularly in Nietzsche, Freud and Arendt. This signifies an empirical attitude according to which multiple inspirations can be used to develop a sophisticated array of perspectives on a subject of study.

A Critical Perspective on Morality and Violence

Arguably, Désirée's most important guiding principle for military ethics was that any productive discussion on military practice and violence starts by avoiding moralization: 'ethics is not political correctness'. This guiding principle is inspired by the views that Nietzsche and Freud developed on the nature of morality and its connections with violence and aggression. In Nietzsche's case, this was developed out of a critique on Christianity, while Freud developed his views out of a critique on a suffocating Victorian atmosphere in Vienna. Their critical analyses inspired Désirée's view that prevalent moral principles should also become a subject for discussion in moral education. Inspired by Arendt, the essence of moral education is therefore critical thinking, which is particularly important in a military academy focused on the serious business of executing the state's monopoly of force.

How do Nietzsche and Freud view morality and how do they relate morality to violence? Furthermore, why should they be considered particularly relevant in the first place? For a philosopher interested in violence, the obvious reason to be inspired by Freud is his account of the ego as being at the mercy of unconscious impulses. Aggression is one of those impulses, as is Eros. In correspondence with Einstein on 'Why war?', Freud claims the following with respect to aggressive impulses (1933, p. 45): 'With the least of speculative efforts, we are led to conclude that this instinct functions in every living being, striving to work its ruin and to reduce life to its primal state of inert matter'. It is against the background of this postulate of primitive impulses that Freud's ambivalent perspective on morality emerges. Unbridled expression of primitive impulses would make social life pretty much impossible, and, in Freud's scheme, they are kept in check by repression and sublimation. In the same correspondence with Einstein, Freud postulates that violence is a force that is constitutive of society (1933, p. 29):

> Brute force is overcome by union, the allied might of scattered units makes good its right against the isolated giant. Thus we might define 'right' (i.e. law) as the might of a community. Yet it, too, is nothing else than violence, quick to attack whatever individual stands in its path, and it employs the selfsame methods, follows like ends, with but one difference: it is the communal, not individual, violence that has its way.

In this view, the rule of law rests on violence, as it depends upon the collective force to overpower individual outbursts of primitive violent impulses. The Super Ego or moral conscience occupies a crucial position in these processes, but as Freud also explains to Einstein, moral conscience has an aggressive quality that is turned inwards. Before Freud, in *Ecce Homo* Nietzsche claimed about conscience that (2007, p. 79): '[...] it is the instinct of cruelty, which turns inwards once it is unable to discharge itself outwardly. Cruelty is here exposed, for the first time, as one of the oldest and most indispensable elements in the foundation of culture'. Whereas Nietzsche worked in the latter stage of his active life towards possible ways to re-evaluate values, and so to prevent nihilism, the Freudian position remains fundamentally ambivalent (Rieff, 1979, p. 343):

> Happiness can never be achieved by the panaceas of social permissiveness or sexual plentitude. Order can never be achieved by social suppression or moral rigor. We are not unhappy because we are frustrated, Freud implies; we are frustrated because we are, first of all, unhappy combinations of conflicting desires. Civilization can, at best, reach a balance of discontents.

For Désirée, it has never been about whether aggressive impulses—or the opposite forces of Eros—are to be regarded as innately biological, a consequence of socialization or an entirely social construct. Her agenda was not focused on developing a particular philosophical anthropology. The point for her is the empirical observation that individuals, beneath a veneer of espoused moral excellence, are far less civilized than they seem, or, rather, that individuals embody both peaceable and belligerent forces. And that is a point that matters for a military organization tasked with executing the state's monopoly of force. Understanding how psychosocial dynamics influence the situations in which 'the monopoly of force is applied' is therefore crucial for the military profession in terms of understanding the dynamics within military organizations themselves and within the environments in which they operate. This is psychosocial in the sense that it requires an individual-focused understanding of how certain tensions may induce violent behaviour, as well as an understanding of such dynamics in groups, organizations and society.

Nihilism and Moral Judgment

A typical initial reaction that Désirée would encounter in workshops for military personnel who had been deployed on missions in Afghanistan, Iraq or Bosnia was scepticism and resistance. They had expected the professor in military ethics to point out their moral shortcomings and hold moralistic lectures about proper behaviour. However, by focusing on their experiences, on the dilemmas they experienced and their struggles in confronting those dilemmas, the atmosphere would soon turn around. In other workshops, Désirée would reflect on 'the enduring appeals of battle', of which 'the delight in destruction', postulated by Glenn Gray (1998), is one which typically fascinates students. However, as valid as it might seem from an academic point of view, and as effective as it might be for capturing the attention of students, one might ask if it is actually sensible to promote critical reflection on morality in a military organization that is involved in the serious business of executing the state's monopoly of force. Notwithstanding the fascination of students, it could be asserted that critically discussing the ethical principles underlying military deployment might undermine the very basis for its legitimacy.

This relates to the important issue of nihilism. One might believe that a critical view on morality and promoting critical reflection can provoke the nihilistic view that dismisses the idea of morality altogether. A nihilistic view on morality might lead to an idea that 'anything goes in the reality of violent conflict' and that is 'just the way it is' in military practice. Such a perspective would be disastrous for an institution such as a military that is founded upon legitimacy. Désirée took up the theme of nihilism in her inaugural lecture (2008). She discussed *Heart of Darkness*, a novella by Joseph Conrad that was the basis for the movie *Apocalypse Now* (1979) directed by Francis Ford Coppola. Central to the movie is a Colonel Kurtz, who experiences great internal conflict in the course of his tour. The theme that Désirée picked out of both works is that of morality being experienced as a burden—as something that makes one powerless—in violent conflict, which in Colonel Kurtz's case led to the desire to be rid of it. Désirée asked her audience if the feelings of Kurtz might be understandable: 'are we not better off without moral judgment?' To answer this question, she turned to Hannah Arendt, who was fascinated by the question as to why some rejected the morality of the Nazis and some did not. Arendt's straightforward answer was that those who did kept thinking for themselves (2003, p. 31):

> The precondition for this kind of judging is not a highly developed intelligence or sophistication in moral matters, but rather the disposition to live together explicitly with oneself, to have intercourse with oneself, that is, to be engaged in that silent dialogue between me and myself which, since Socrates and Plato, we usually call thinking. [...] In this respect, the total moral collapse of respectable society during the Hitler regime may teach us that under such circumstances those who cherish values and hold fast to moral norms and standards are not reliable: we now know that moral norms and standards can be changed overnight, and that all that then will be left is the mere habit of holding fast to something. Much more reliable will be the doubters and skeptics, not because skepticism is good or doubting wholesome, but because they are used to examine things and to make up their own minds.

This quote indicates that Arendt shares the critique on civilized morality. Arendt emphasizes that one of the gruesome facts of the Nazi terrors was that they occurred within a legal framework. Moral and legal frameworks may therefore not only repress aggressive impulses but also legitimize violence. Questioning existing norms and standards, fortunately, will not automatically release our repressed impulses. Instead, it will help to critically examine the relationship between our norms, standards and impulses. In her inaugural lecture, Désirée emphasized that the kind of dialogue Arendt advocated actually prevents rather than creates the nihilism of Colonel Kurtz. As Désirée stated in *The Dark Side of Obedience* (Verweij, 2004, pp. 156–157):

> We need soldiers and officers who can obey in the Nietzschean sense, that is to say, who have developed inner discipline, which implies that they are also able to obey to themselves, that they can listen to themselves, in the way Socrates described. The inner dialogue, the ability to think, can confront them with the virtues - of which compassion is one - which they have been taught to cherish. This implies that they will know when and how to obey and at the same time be compassionate, and in acting that way, they will put into practise the precise intention of the code of conduct.

In Désirée's view, such an attitude is the best possible path to moral judgement, also in adverse conditions.

Violence and Truth

Désirée's guiding principle that any productive discussion on military ethics starts by avoiding moralization implies that ideas that at first instance appear counterintuitive are not just taken into account but also might trigger important insights and might shed light on particularly significant dynamics. As the discussion above reveals, this principle is related to Nietzsche, Freud and Arendt who, despite differences, come together in a critique on civilized morality and an emphasis on the central importance of critical thinking and honesty. An example of a counterintuitive idea with generative potential is that of a particular kind of 'truth' that can be found in the confrontation with violence. This particular idea emerges in Freud's *Reflections on War and Death*, written after World War I. Establishing that for Freud moral inflation induces moral depression, Rieff (1979, pp. 312–313) describes this theme as follows:

> Though war may seem to the cultured as a "regression", some regressions may be therapeutic. War drew away the superficies of culture; it "has the advantage of taking the truth more into account, and of making life more tolerable for us again." War and revolution (they amount to the same thing, for Freud, since both have this regressive character) were natural therapies for the over-civilized as psychoanalysis was an artificial one (quote in original).

Freud signifies that aggressive impulses not only emerge when repressions and sublimations fail to function properly, they may also emerge as the result of repression itself. Rieff calls this therapeutic aspect of war an eccentric and apocryphal theme in Freud's work, but his critique on culture is certainly not.

While to some the idea of a kind of truth that can be found in the confrontation with violence might seem far-fetched at best, there is a particular connection between this perspective and contemporary perspectives on military trauma. Robert J. Lifton (1973) pointed out that many veterans were disillusioned and shocked by how moral authorities reacted to their reflections on their experiences in battle. When they confronted moral authorities (officers, priests, psychologists) with doubts about the things that they had done ('Am I a war criminal?'), they found that these authorities strongly rejected their questions ('You are a hero!'). Lifton observed that instead of experiencing such counterstatements as uplifting, veterans could experience them as disillusioning. They began to question both the moral quality of these authorities and the broader culture that they represented. To the veterans, they revealed a 'counterfeit universe'—a universe of moral betrayal at the core of civilized morality. In that sense, there is a particular kind of truth to be found in the confrontation with violence. Lifton observed a particular sensitivity to hypocrisy in veterans. The idea that the confrontation with violence might reveal a kind of 'truth' also emerges in contemporary research on moral injury (Bica, 1999; Molendijk, 2021) and has been an important principle in the research programme that Désirée established.

The foregoing indicates that the particular 'truth' that the confrontation with violence may reveal further underlines the importance of talking about violence. The confrontation with the brutal reality of military deployment provides important input that can help to either validate or criticize moral and legal frameworks that were used to legitimize the use of force. Therefore, talking about violence is a key aspect in the legitimate use of force, while at the same time it is often the most controversial one.

Bureaucracy and the Banality of Evil

The previous discussion relates to a further relevant perspective on violence. This is the influence of bureaucracy, and this explains Désirée's interest, perhaps a somewhat unconventional one for a philosopher, in organizational structure. Arendt's theme of 'the banality of evil' and its relationship to totalitarian tendencies in bureaucracies constitutes Désirée's main inspiration with regard to this issue. What triggered the idea of 'the banality of evil' was Arendt's observation that behind one of the main perpetrators of the Holocaust, Adolf Eichmann, was not a manifest monster but an appallingly average man. She saw Eichmann, whose trial in Jerusalem she attended, as a bureaucratic operative that concentrated on following the rules. Realizing she had implicitly expected that Eichmann's 'evil' would expose itself as overtly monstrous, she found the idea that 'evil' might be banal even more shocking.

This led her to criticize bureaucracy as an organizational form that creates 'a rule by nobody'. In *On Violence* (1970, p. 80), Arendt discussed the tyrannical characteristics of bureaucracy:

> the greater the bureaucratization of public life, the greater will be the attraction of violence. In a fully developed bureaucracy there is nobody left with whom one can argue, to whom one can present grievances, on whom the pressures of power can be exerted. Bureaucracy is the form of government in which everybody is deprived of political freedom, of the power

> to act; for the rule by Nobody is not no-rule, and where all are equally powerless we have a tyranny without a tyrant.

The idea Arendt expresses here is that bureaucratic rule by nobody can produce tyrannically violent results as an aggregate effect of rule-following agents who operate as cogs in a machine and are alienated from the very processes they contribute to. Not coincidentally, Ulrich Beck (1995) called bureaucracies 'systems of organized irresponsibility'. Bauman (1989) took the critique on bureaucracy even further. His position is explained by De Swaan (2015, p. 41) as follows:

> according to Bauman, the civilizing process, in its drive toward evermore pervasive rationality, is essentially a two-sided phenomenon. It not only promotes humane and lawful modes of social existence, but also facilitates the "rationalization" of the unrestrained use of violence, devoid of any moral calculus or ethical inhibition.

Désirée's critical perspective on ethics in military organizations takes into account the risk of the rationalization of immoralities. At the same time, she is careful not to overstate this point. As an interdisciplinary-minded philosopher, she takes critiques that reveal that Arendt's view tends to overlook other important factors involved in the production of 'evil' seriously. Mandel (2002, p. 279) considers the banality perspective an oversimplified situationist account; that is, it overstates the significance of the influence of the immediate environment. Furthermore, the idea that Eichmann was a mere banal rule follower was criticized by Stangneth (2014) on the basis of historical evidence. In fact, Eichmann had been quite a fanatical Nazi. Similarly, De Swaan (2015) emphasized that, being a member of the top of the Nazi hierarchy, Eichmann would have been the opposite of a banal rule follower. However, while he shares the critique on oversimplified all-explaining situationist accounts, he does emphasize the importance of situational conditions for explaining violent behaviour and he is not prepared to put aside the influence of bureaucratic structures (2015, pp. 22–23):

> Arendt's thesis on the "banality of evil" does not stand critical scrutiny, certainly not as applied to Adolf Eichmann or other Nazi leaders, nor for that matter to the rank-and-file killers. Her model might, however, fit the countless minor middlemen of the Holocaust: the administrators in the civil registry who passed on the names of the prospective victims, the local police who rounded them up, the engineers who transported them in cattle cars, the contractors who built the gas chambers and supplied the extermination camps [...] most of them, indeed, were in some sense banal.

This insight poses a major problem for the armed forces. The bureaucratic character of the military organization as one of the organizations that executes the state's monopoly of force is the very foundation of its legitimacy. The critique on the potential corruption of this organizational form is therefore both highly significant and deeply problematic for the military. This point about the dangers of 'the rule by nobody' can be connected to contemporary discussions about autonomous weapons. Emphasis is placed upon the importance of establishing 'meaningful human control' in such weapon systems (Ekelhof, 2019). Yet, Arendt's views about bureaucracy indicate that there is something terrifying about such control when it is organized in bureaucratic systems that are 'ruled by nobody', while the nature of the weapon systems themselves may lead to the lack of an embodied military presence at the locations of violence.

The instrumental character of a bureaucratic system encourages the dressing up of violence in technical instrumental language, and the language of moral responsibility can neatly serve such technocratic purposes. This last issue, the use of the language of morality for violent purposes, is the subject of Désirée's proposed concept of 'moresfare', which she discusses in a chapter in this book. For a military organization whose operatives may encounter situations in which they are required to apply violent force, an understanding of the dynamics of bureaucracy is vital. The first step in gaining that understanding is to talk about violence.

The Theme of This Book and the Overview of the Chapters

Understanding the multifaceted and partially hidden dynamics of destructive and violent behaviour is essential for a military organization. Bringing together a variety of expertise, this volume reflects on confrontations with violence in extreme conditions and the various challenges resulting from such confrontations. This volume reflects on this theme from a variety of disciplines such as philosophy, psychology, anthropology, political science, and organizational studies. The contributions are clustered in three themes.

The first theme is predominantly conceptual and focuses on **multi—nd interdisciplinary reflections on violence and military ethics.** The contributions to this theme aim at coming to grips with what confrontations with violence mean. The first contribution in this theme is by **Paul van Tongeren** and is entitled **Language and violence.** In this chapter, Van Tongeren argues that thinking and speaking about violence is a paradoxical affair. Violence is not so much something we can think about, but something we have always been against and therefore actually think against. On the basis of three short texts by three contemporary French authors, this chapter provides an impetus to answer the unanswerable question of what violence actually is in service of the unachievable task of overcoming violence. The second contribution is by **Tine Molendijk** and is entitled **Military trauma and the conflicted human condition: Moral injury as a window into violence, human nature and military ethics.** In this chapter, Molendijk contends that 'moral injury', which refers to the lasting psychological impact of morally critical situations and has been called the 'signature wound' of contemporary military operations, is characterized by dynamic moral complexity. This chapter examines morally injurious conflict at the psychological micro level in relation to structural tensions at the level of the organization, politics and society. In doing so, it conceptualizes moral conflict as inherent to being human and as something that can manifest itself in destructive ways in extreme contexts. The third contribution to this theme is by **Eric-Hans Kramer, Max Visser and Matthijs Moorkamp** and is entitled **Exploring the relevance of the systems psychodynamic approach to military organizations**. In this chapter, they focus on anxiety that confrontations with violence may cause in operators and the implications for military organizations. Rules, procedures and working practices in organizations may provide important defences against anxiety that arises when doing risky and dangerous work. Such defences are often institutionalized in such a way that they are taken for granted. Given that confrontations with violence are at the core of the military profession and given that such confrontations can be anxiety provoking and emotionally intense, this chapter explores the potential value of the systems psychodynamic perspective for understanding structures and processes in the military organization. An important idea in this perspective is that while defences against anxiety may have important primary functions, they may have secondary dysfunctional effects in a changing world with changing professions.

The second theme focuses on **recent cases and developments.** The first contribution in this theme is by **Erella Grassiani** and focuses on the Israeli Army and moral reasoning. It is entitled **Instrumental morality under a gaze: Israeli soldiers' reasoning on doing "good".** This contribution analyzes soldiers as violent actors; it shows different strategies of legitimization used by soldiers when explaining their use of force, which is fluid and prone to change according to context. It also shows how soldiers use 'instrumental morality' to explain their use of force and make it legitimate, even if it falls outside of the formal instructions of the military. The second contribution in this theme is by **Teun Eikenaar** and is entitled **Soldiers as street level bureaucrats?** In this contribution, Eikenaar argues that military organizations and military professions have changed in considerable ways over the last thirty years. Among other things, changes in mission outlook, new professional identities, and development towards what some call the 'constabularization of the military' have important consequences for the discretionary autonomy of military personnel. In a sense, this has meant that many military professionals are increasingly expected to work in circumstances that are more reminiscent of police work than of stereotypical 'command and control' settings. In these settings, having to decide on when and how to apply coercion implies specific (and maybe unexpected) moral demands and expectations. This contribution compares military and police force front-line workers and the implications that recent military developments might have for the moral demands on front-line workers. The third contribution in this theme is by **Marenne Jansen and Eric-Hans Kramer** and is entitled **Does the comprehensive approach have any future as a strategy for intervention?** This chapter focuses on post-conflict settings and starts from the observation that the first decade of the new millennium saw several international interventions based on a firm belief in the nexus between security and development, which contends that security and development are interconnected. The tragic events in Kabul in August 2021 did not only mark the end of the US mission and twenty years of 'western' military intervention in Afghanistan, but arguably also marked the end of the 'comprehensive approach' as a functional model for foreign intervention, integrating security and development as a peacebuilding strategy. This chapter evaluates the application of the comprehensive approach of the last decades by connecting it to three perspectives on the human security concept. By doing so, it analyzes the validity of the comprehensive approach as an intervention strategy.

The third theme comprises **a number of answers to current challenges** that military organizations have developed in response to the challenges involved in the confrontation with violence. The first contribution to this theme was written by **Lonneke Peperkamp and Nikolaus Braun** and is entitled **Contemporary just war thinking and military education.** Starting from the observation that armed forces operate in extreme circumstances, where they may be confronted with violence and/or use force themselves, they focus on just war theory which provides ethical guidance on how to deal with such violence. There is an intense discussion within this field on the appropriate goal of just war theory, on how to provide that guidance and on the relationship between morality and international law. This chapter analyzes the three main schools in contemporary just war theory and defends a practically oriented account. The second contribution is by **Peter Olsthoorn** and focuses on the topic of **Educating for restraint**. This chapter reflects on several significant challenges regarding ethics education in the military. What should the basis of military ethics education be and what should the goal be? Who provides the education, and how? How do you reconcile the practical goals of ethics training with academic education? And what do we know about the effectiveness of that ethics education? The third contribution in this theme is by Edgar Karssing and is entitled **The e-word (emotions) in military ethics education: Making use of the dual-process model of moral psychology.** It starts from the contention that dual-process theories within the field of moral psychology state that two sorts of processes can be differentiated: an affective, associative process and an analytical, rule-based process. These theories can help us to address the importance of emotions and feelings in moral judgement and decision-making. Emotions could lead to destructive behaviour in a violent context. However, it is possible to cultivate emotions so that people will act in a responsible way when faced with extreme conditions. What can we learn from dual-process theories for teaching military ethics? The fourth contribution to this theme is entitled **The Dutch Approach to Ethics: Integrity Management in the Military** and was written by **Miriam de Graaff and Claire Zalm.** In this contribution, they address the pillars of the integrity management system of the Netherlands armed forces in order to provide background on integrity management for public managers who currently work in a high-stakes environment, such as the military. They intend to open a discussion within the field of military ethics on how integrity management can be carried out in environments that are highly demanding, using the Dutch military as an example.

In the **epilogue,** first **Désirée Verweij** provides her contribution. Rather than looking back, she brings forward a new concept 'moresfare' in order to reflect on the current trends that she detects in the realm of hybrid warfare. In her contribution, entitled **'Moresfare' and the resilience paradox: Ethics as the terra incognita of hybrid warfare and its challenges**, she states that hybrid warfare not only poses a challenge for ethics; it is also a conscious use (and often misuse) of ethics. In hybrid warfare, values and the emotions connected to these values are played on, influenced, questioned, and moulded in such a subtle way that the target audience hardly realizes what is actually happening; this can be typified as 'moresfare'. The chapter analyzes the often shielded form of 'moresfare' and its connection to the more familiar concept of 'lawfare'. In doing so, it also addresses the concept of resilience, which is often believed to counteract the hybrid use of force and thus the effects of moresfare. In the second part of the epilogue **Tine Molendijk and Eric-Hans Kramer** formulate **Concluding reflections**. This concluding chapter draws together some of the key themes and insights from the contributions. It discusses how some characteristics of military intervention and views on military ethics have changed over the decades, and how some things have remained the same and goes on to reflect on the implications thereof. On the basis of this discussion, it proposes recommended avenues for future research, policy, and education.

As we stated at the beginning of this introduction, interdisciplinary explorations of the topic of 'the confrontation with violence' have been at the core of the work of Désirée Verweij. The various contributions in this volume indicate that violence relates to many aspects of the everyday reality of military organizations. Désirée's remark from a while back that at the military academy 'nobody ever talks about violence' has initiated discussion on a phenomenon that is intrinsically linked to the military organization. This volume does talk about violence, and the various contributions indicate why this is a topic that we do not intend to stop talking about in the future.

References

Arendt, H. (2003). *Responsibility and judgment* (Edited by and with an introduction by Jerome Kohn). Schocken Books.

Arendt, H. (1970). *On violence*. Houghton Mifflin Harcourt Publishing Company.

Bauman, Z. (1989). *Modernity and the Holocaust*. Polity Press.

Beck, U. (1995). *Ecological politics in an age of risk*. Polity Press.

Bica, C. C. (1999). A therapeutic application of philosophy. The moral casualties of war: Understanding the experience. *International Journal of Applied Philosophy, 13*(1), 81–92.

Conrad, J. (1995). *Heart of darkness with the Congo Diary* (Edited and with an introduction and notes by Robert Hampson). Penguin Books.

Einstein, A., & Freud, S. (1933). *Why war?* International Institute of Intellectual Co-operation. League of Nations.

Ekelhof, M.A.C. (2019). Moving beyond semantics on autonomous weapons: Meaningful human control in operation. *Global Policy*, 1–6.

Freud, S. (2013). *Reflections on war and death*. Redditch: Read Books, Ltd., originally published in 1924, Internationaler Psychoanalytischer Verlag.

Glenn Gray, J. (1998). *The warriors: Reflections on men in battle*. University of Nebraska Press, first edition 1959.

Lifton, R. J. (1973). *Home from the war: Learning from Vietnam Veterans*. Other Press.

Mandel, D. R. (2002). Instigators of genocide: Examining Hitler from a social-psychological perspective. In L. S. Newman & R. Erber (Eds.), *Understanding genocide: The social psychology of the Holocaust* (pp. 259–284). Oxford University Press.

Molendijk, T. (2021). *Moral injury and soldiers in conflict: Political practices and public perceptions*. Routledge.

Nietzsche, F. (2007). *Ecce Homo*. Oxford University Press.

Rieff, P. (1979). *Freud: The mind of the moralist*. University of Chicago Press, third edition, first edition 1959.

Stangneth, B. (2014). *Eichmann vor Jerusalem: Das unbehelligte Leben eines Massenmörders.* Rowohlt Verlag.

De., Swaan, & A. (2015). *The killing compartments: The mentality of mass murder*. Yale University Press.

Verweij, D. (1993). *Ariadne en Dionysos: Vrouw-metaforen en verlangen in het werk van Nietzsche.* Thesis publishers.

Verweij, D. (2004). The dark side of obedience: The consequences of Hannah Arendt's analysis of the Eichmann case. In: *Professional Ethics, 6*(1), 19–40.

Verweij, D. E. M. (2007). Military ethics: A contradiction in terms? In J. Toiskallio (Ed.), *Ethical education in the military: What, How and why in the 21st century?* (pp. 43–62). National Defence University.

Verweij, D. (1999). Nietzsche. In: Studium Generale. *Grote Filosofen.* Uitgeverij Nieuwezijds: Amsterdam.

Verweij, D. (2008). *Denken in dialoog. Ethiek en de militaire praktijk.* Nederlandse Defensie Academie.

Eric-Hans Kramer is Professor in Military Management and Organization Studies at the Netherlands Defence Academy. Before that, he was associate professor Human Factors and Systems Safety at the same organization. He is involved in several research projects into organizational issues the military organization is being confronted with and with the development of research and educational programs into the psychosocial dynamics in military organizations.

Tine Molendijk is an interdisciplinary-minded cultural anthropologist specialized in the topics of violence, military culture and mental health, in particular post-traumatic stress disorder (PTSD) and moral injury. Her PhD research focused on the role of political practices and public perceptions in moral injury in Bosnia and Afghanistan veterans. Currently, she works as Assistant Professor at the Faculty of Military Sciences at the Netherlands Defence Academy (NLDA) and as a Research Fellow affiliated to the Radboud University. Among her publications is the award-winning book Moral Injury and Soldiers in Conflict: Political Practices and Public Perceptions (Routledge, 2021).

Multi—and Interdisciplinary Reflections on Violence and Military Ethics

Language and Violence

Paul van Tongeren

Not that long ago, there was a war in Yugoslavia. NATO was using considerable force in an attempt to prevent Serbia from pursuing its ethnic policy in Kosovo. This use of violence was justified on humanitarian grounds after political and diplomatic pressure had been applied for many months in an attempt to persuade the regime in Belgrade to change its position. However, the talks finally broke down and NATO turned to violence, since actions speak louder than words.

With regard to what we refer to as 'senseless violence', often street violence targeting innocent citizens, a notable factor is that this is often provoked when the victim calls the perpetrator to account. Joes Kloppenburg, Meindert Tjoelker and Kerwin Duinmeijer are just three high-profile names among the many dozens of anonymous victims of senseless violence in the Netherlands in the 90 s. Joes, Meindert and Kerwin, and possibly many more of them, were attacked and murdered by people whom they had confronted about their conduct.

These are two examples of violence that affects us deeply, and both examples suggest a link between violence and language, albeit in a contrasting manner. Philosophers work with words. As we discuss and contemplate the theme of violence, we probably feel a greater sense of powerlessness than we do when discussing other themes. This powerlessness is probably due to the nature of the theme of violence and the nature of the activity with which we intend to address it, namely thought and speech—in other words, language.

It has been pointed out that thinking and speaking about violence is paradoxical. This is true for at least two reasons. First, violence, like evil in general, appears to be characterised by absurdity and irrationality. If violence is characterised by its

This is a translation of a contribution originally published in Dutch. Original Source: Paul van Tongeren, Taal en geweld. Uit boek: Koen Boey (red.), Filosoferen over geweld. Acco, Leuven/Leusden 2000, pp. 13–30.

P. van Tongeren (✉)
Department of Philosophy, Radboud University, Nijmegen, The Netherlands
e-mail: pvtongeren@mac.com

E.-H. Kramer and T. Molendijk (eds.), *Violence in Extreme Conditions*,
https://doi.org/10.1007/978-3-031-16119-3_2

confliction with rationality, it will be difficult to comprehend—in other words, the more we think we understand it, the more elusive it may become. The second reason is even more persuasive. As C. Verhoeven and others have often pointed out, violence is not a subject that we can contemplate freely, something that we can think *about*, but something that we are *against* and thus think *against* (Verhoeven, 1967). While we normally look *towards* the phenomenon under consideration, in the case of violence we seem to look *away* from it. We are therefore strongly prejudiced, or to put it more kindly, we have a strong engagement.

Violence seems primarily to demand that we combat it rather than reflect on it. But how can we be sure that our battle against violence is not in itself violent, or whether in fact it needs to be in order to succeed? And what does that imply for our engagement? What does it mean if we have a violent urge to combat violence? What is violence if it is both the target and the weapon? However engaged we may be, we must not shy away from reflecting on what violence actually is. More importantly, we must avoid confusing the question of what violence is and why people feel compelled to use it with the question of what we can do to combat it.

Perhaps we can easily resolve the paradox of using violence to combat violence by distinguishing between different types of violence. We can distinguish the violence of war and senseless street violence, but also the violence of education. In Plato's allegory of the cave, which describes the path along which humankind can escape from the prejudices and false beliefs of prevailing opinion, it is notable that he frequently uses terms that elucidate how violent this liberation would be. True knowledge seemingly has a need of violence. Besides the violence of education and the physical violence of battle, we can distinguish other forms of violence, such as the devastation of overpowering experience or the destructive power of the natural world. This implies a second warning that must be heeded if we seek to contemplate violence: we should not readily assume that that the word violence is always being used to refer to the same concept. At the same time, scientific or pseudoscientific definitions of all the different forms of violence and their preconditions will not suffice. In other words, we cannot avoid asking the philosophical question of whether there is an all-inclusive notion that encompasses all forms of violence.

The more pervasive a concept is, the more difficult it is to define, since definition is always a limiting factor. *Omnis determinatio est negatio.* Anyone aiming to define something as all-encompassing as violence will have to search for a contrasting concept with which to delimit it. This is where language comes in. Violence and language (the instrument of reason and logos) appear to be diametrically opposed and therefore to define each other, which is why we often encounter this opposition in the history of the philosophy of violence. Incidentally this position reinforces the paradox mentioned above. To the extent to which language and violence are opposites, if indeed they are opposites, we will always distance ourselves from violence simply by speaking, as a result of which violence will always remain dumb, inaccessible to words, and we will never resolve the paradox. However, this cannot be a reason to refrain from contemplation. On the contrary, philosophers have tended to devote the closest and most constructive attention to those questions to which they could have known in advance (and usually did) that answers could not be found.

This introduction sets the scene for what I would like to do next. Based on three short texts by three contemporary French authors, I aim to explore several avenues in pursuit of answers to the unanswerable question of what violence actually is, in a quest to attain the unattainable and conquer violence. These three authors, in chronological order, are Eric Weil, Emmanuel Levinas and Paul Ricoeur, who will be discussed here in reverse order. I have resisted the temptation to add Jacques Derrida, with his essay *Violence et métaphysique* (1964), as a fourth author. I will not be guided solely by their writing but will analyse the texts on the basis of a framework that sprang to mind while reading their work. I will therefore simplify their writing to fit this structure. I will use them freely for my own story, interpret them in my own words, and in doing so will subject them to violence. So be it.

Language Versus Violence

Over 30 years ago Paul Ricoeur wrote an introduction to a discussion on violence, which, together with an article by Eric Weil, was published in 1967 entitled *Violence et Langage* (Ricoeur, 1967). It forms a useful starting point for our argument, firstly because Ricoeur explicitly thematises the general concept of violence, secondly because he does this in the context of the opposition between violence and language, and thirdly because he explicitly addresses the practicality of a philosophy of violence.

Ricoeur considers it to be the philosopher's task to define the concept of violence in such a way that all forms of violence are incorporated in that definition. Although some will object that this precludes discussion of specific concrete problems (in the 1960s Ricoeur would have been thinking about the violence of repression and revolution, whereas nowadays humanitarian intervention and senseless street violence tend to be more topical), in order to discuss such problems it is ultimately important to know what we mean when we refer to them in terms of violence. The common denominator is the violent character of widely differing phenomena such as the violence of nature—as embodied by a hurricane—on the one hand and the interpersonal violence of murder on the other, as well as the many other forms of human violence that lie in between.

Before I present Ricoeur's answer to that question I should like to make another point. It is important to recognise that almost all human violence falls in the intermediate zone between the two extremes. In order to understand this we must realise that the violence of nature is found not only in phenomena such as hurricanes, floods and volcanic eruptions but also in our own violent nature, in emotions such as desire, fear and hatred. Who has mastery of their own violent nature? Who can guarantee that any violence they exercise will remain within the bounds of their own intentions and will have no effect other than that intended? The fact that violence always retains some of its origin as a force of nature puts the role of intention into perspective if this origin is called upon to legitimise the use of violence. Human violence is always a combination of the violence exercised and the violence experienced, both internally and externally.

This further complicates the question of what connects such widely differing concepts. What is the common thread unifying the different forms of violence? Ricoeur's answer is that this unification lies in the opposition between violence and language: '*Ce qui fait l'unité de l'empire de la violence, c'est qu'il a le langage pour vis-à-vis*' (1967, p. 87). Along the entire line, from one end to the other, the two concepts adjoin '*comme deux contraires exactement ajustés chacun à l'extension entière de l'autre*' (id.). Here Ricoeur is referring to language not in the sense of linguistic structure comprising vocabulary and grammar *(langage)* but rather in the form of discourse (*discours*) and ultimately rational argument. Where there is discourse, in the extent to which discourse is used, there is no violence, and where violence is used, in the extent to which violence is exerted, there is no discourse. An argument for violence is therefore paradoxical, and we are struck dumb by savage violence. The aim of violence in its pure form is to silence and objectify the other.

This delineation between language and violence provides some clarification in relation to the previous question of how to unify all the widely differing forms of violence. Even the violence of nature only exists as such in the mind of man, who is left speechless. Humans who lose themselves in violence may be labelled as bestial, but that does not necessarily mean that animals acting in a similar way can be labelled as violent. Pure violence may be associated with the absence of language, but that absence must actually be discernible. If language is completely missing, and the contrast with language cannot be made because there is no linguistic expression, there can be no question of violence either. It is the opposition with language that makes violence what it is. The notion of a discernible absence introduces a dialectic that is discussed in more detail in the work of Eric Weil in particular. Ricoeur does not enter into this discussion in any depth, but rather shows that this absolute opposition between language and violence is, of course, an abstract concept.

In the real world, the manifestations are almost always in hybrid form. Human violence is always interwoven with language and human discourse is always interwoven with violence. Ricoeur examines the latter observation in three different spheres: the world of politics, where this phenomenon is self-evident, the sphere of poetry, where it is seemingly improbable and the realm of philosophy. In all of these spheres, he underlines the intermingling of what he calls *expression* and *sens,* albeit in highly varying proportions. The fact that a person seeks to express himself and, whether issuing orders or evoking ideas, always acts within assumed frameworks using defining words alludes to the violence inherent in every form of expression, but since this person also seeks to impart information and offer clarification with his words, this remains discourse and is even the voice of reason in every instance of violence.

But what is the purpose of constructing an absolute opposition between language and violence only to subsequently acknowledge that actually only hybrid forms exist, and that all violence contains an element of reason and all discourse retains an element of violence? Ricoeur's answer is that this absolute opposition lays the foundation for establishing an ethics of violence, or rather how to handle violence. The absolute opposition between violence and discourse provides a basis for answering the practical question of how we should deal with the inevitable reality of violence. These ethics are summarised by Ricoeur in three *modestes règles*:

1. The first rule is simply to be aware of the absolute opposition. It is important to adhere to the essential principle that violence and language are polar opposites, even though this principle may be formal and still hollow. This principle must be acknowledged in order to recognise violence when it occurs but also to resort to violence if there is no other option. However, this allows violence to be concealed, to be defended as if it belonged in the realm of reason and dialogue. Whoever resorts to violence—and Ricoeur acknowledges that this is sometimes unavoidable—must accept a certain level of culpability, even though according to Ricoeur this may be limited (*culpabilité limitée*). After all, calling a crime a crime is the first step on the road to salvation.
2. The second rule is to apply the non-violent nature of discourse as an imperative rather than merely as a formal truth—not out of blindness to the inevitability of violence or to the violence inherent in discourse, but in order to prevent violence from ever becoming total. '"Thou shalt not kill" is always true even when it is not applicable' (Ricoeur, 1967, p. 93). The advocacy of non-violence is always worthwhile in a world in which violence is sometimes necessary. It occupies a meaningful position in the dialectic between inclination and responsibility.
3. The third rule is to attempt to banish violence from discourse as far as possible by respecting the plurality of language and manners of speaking. There is the language of calculating reason, but also that of totalising reason, prophetic appeal and mythical invocation.

Ricoeur's answer to the question of what we should do to combat violence is pervaded by the realisation that violence is unavoidable, but that is precisely why it is so important to make every effort to adhere to the formal and idealistic division and the opposition between violence and language.

Elementary Discourse Versus Totality Of Violence

In the first place it is primarily the necessity and universality of violence that Emmanuel Levinas emphasises in his article. First published in 1953, entitled *Liberté et commandement* (Freedom and Command in the English translation) (Levinas, 1987, pp. 15–23). He introduces violence through an analysis of the act itself. An act can be defined firstly in terms of the degree to which one can say that an individual is acting *on his own agency*, the degree to which he is not the helpless plaything of external forces but is rather the origin of those forces, and secondly in terms of the extent to which an individual *actually effectuates something*—in other words, he does not keep his intentions within himself but puts them into effect in situations in the outside world. In the latter case, the more resistance offered by the outside world, the more powerful the effect. Actions are more powerful if the agent imposes his will on a stubborn stone and achieves the desired result than if he meets with no resistance and simply gives a gentle push to an already rolling stone. It is therefore not necessarily a hard object such as a stone that offers the greatest resistance but rather an active entity that seeks to exert its own influence: another freedom. I must therefore conclude that an act in the strongest sense of the word can be considered to be a command, or even subordination, or violence. War is the normal state of affairs for beings that engage in deliberate acts (in other words: mankind). Violence is not simply the prerogative of mankind, it also characterises us, since it is the clearest and most powerful expression of humanity in the sense of having freedom and the ability to perform deliberate acts.

The problem this presents is the problem of political ideology since the modern age. Up to now this analysis by Levinas is not original in any way. Hobbes's statement that a man is a wolf to another man (*homo homini lupus*) makes a similar argument: it is man's humanity, or in this case this particular aspect of his humanity, that makes him a wolf to another man. The other side of man's humanity, namely reason, has been propounded as a counterweight to this unavoidable violence, not only since the time of Hobbes but indeed since Plato. The radical nature of Levinas's analysis lies in his suspicion of all these proposed remedies for violence.

The first of these is the proposition that violence is ultimately ineffective. Although it is possible to force another person to do one's bidding, the other always has the option of continually refusing and in that sense putting up continued resistance. And if he is killed for resisting, his refusal perpetuates. Murdered opponents are often a killer's most persistent persecutors. Against this proposition, Levinas first argues that many tyrants are not too concerned about this implied limitation of their power and that victims of tyrannical violence rarely reap any benefits from their indomitable defiance. Their unremitting freedom to refuse is no more than an awareness of their subjugation. The victims of senseless street violence in our society also refused to submit and that was their undoing. Perhaps they are still announcing their refusal through their public profile but we only know the names of a handful of victims, and many more have died in vain. In addition, it remains to be seen how long we will continue to remember those few names, what effect this memory will have, and how fragile the memory will be. What remains of the powerful protest that was expressed during the White March in Belgium after serial killer and child molester Marc Dutroux was arrested?

Levinas's second argument against this proposition is even more radical. He points out that violence can even overcome this obstacle of resistance. This is where language explicitly comes to the fore. Violence can make use of seductive language and can violate freedom without giving the impression of doing so. Many forms of structural violence probably work this way. Every day we are forced to adopt all kinds of patterns, partly through the advertising and media that control our lives, while being under the illusion that we are choosing the patterns ourselves. Real violence works through language, brainwashing the victim into believing that he actually wants what is being forced upon him. The claim by some philosophers that violence is ultimately fruitless is refuted by Levinas, who points out its effectiveness.

The second proposed remedy to combat violence is that of the rationality of rules and laws. Allowing ourselves to be bound by sensible laws is similar to the way in which Odysseus protects himself against the seductive violence of the sirens by having himself tied to the ship's mast before their seductive powers can take effect. True freedom listens to the laws and protects itself against unreasonable violence by subjecting itself to its own rules of reason. However, notwithstanding the paradoxical nature of a freedom that subjugates itself, it is clear that the concrete reality of this self-protection all too easily leads to renewed violence. Odysseus screams at his men to untie him. In the same way, we rarely perceive the laws of the state and the rules of the institutes in which we work as a form of self-protection. It is often quite the opposite, not only because we are victims of our own violent nature which resists the reins of reason, but also because the rules of reason become violent themselves. Civil servants devise new rules aimed at preserving old rules and at protecting themselves as protectors of the rules. Administrators are constrained by an organisational fervour that threatens to destroy that which requires organisation. Although it is possible to espouse the myth that we have all cooperated in establishing the rules that now subjugate us, that myth often utterly fails to convince even the less free spirited. It is all too easy to violently pervert the rationality of the common good, but this rationality in itself is inevitably perceived as being violent.

It therefore appears that Levinas is incorporating language and reason into violence, as if nothing exists outside violence. However, this reveals the paradoxical reality that, by absorbing everything with which it comes into contact, violence actually undermines itself. Violence is constantly focused on overpowering the other but as a consequence the more it succeeds, the more it is eroded. The more comprehensive the conquest of the other, the less significant the conquered. This dynamic is even more clearly evident in the problem of acknowledgement, whereby the extortion of acknowledgement from another actually forfeits any such acknowledgement. An act of violence requires an adversary, but the appearance of the adversary challenges the violent pursuit of power. In reality this paradox manifests itself as an endless progression in which every power seeks a new challenge, a new enemy.

The solution proposed by Levinas is well known. He attempts to point to an alterity that on the one hand is genuine otherness but on the other hand does not exist in the form of a conquerable obstacle, as a force against which one's strength can be gauged. In the words of Levinas, it is 'a being becoming naked, an unqualified substance breaking through its form and presenting a face' (1987, p. 103). And this is where, despite his far more radical confirmation of violence, Levinas reverts again to the opposition between violence and language. Although the face speaks, it is not what it says that provides the counterbalance. Everything that is said can be understood and in that sense can be appropriated. Levinas is referring here to a form of speech that precedes all discussion or consideration. It does not discuss but rather it addresses, and this elementary language is the only way to break down the totality of violence. As a result the opposition between language and violence is not only maintained but, by becoming more extreme, is in danger of losing its practical significance. What does this resistanceless power signify in relation to international conflict among peoples, street violence, or the violent structure of the technological and economic order? Although it is not entirely insignificant, it merely provides a reminder that the universe of violence is not closed. This is not unimportant, but we should set our sights higher. We should aim to acquire a more effective tool for combating violence than merely a prophetic voice.

A Question of Degree

Perhaps this can only be achieved by putting the contradiction between violence and language, or more specifically the exclusivity of this contradiction, into perspective. In order to investigate this possibility, I conclude with an analysis of the previously mentioned text by Eric Weil (1967). Like Ricoeur, Weil takes the opposition between violence and language as a starting point, but in contrast to Ricoeur he places a far clearer emphasis on the connection created by this opposition. Language and violence are conjoined as a result of their opposition. That is why a phenomenon such as violence can only exist for humans, who are the only creatures to have language and logos. Although like ourselves animals are also subjected to what we call the violence of nature, they do not comprehend this as such. This is because, as far as we know, man is the only creature to contemplate the meaning of life, to seek its sense or significance, for which purpose it needs language. However, violence is a senseless phenomenon, and in that sense it is 'dumb'. It is not so much that senseless violence is a particular category of violence, but rather that senselessness is a defining characteristic of violence. But only a creature that possesses language can identify dumbness as being dumb, and only a creature that searches for meaning can acknowledge absurdity. Violence does not exist without man, but does this also imply that man does not exist without violence? Is man violent by definition? And does this mean that the battle against violence is futile? I suspect that Eric Weil ultimately answers that question in the affirmative but without sounding cynical. For the purpose of clarification I follow the development set out by Weil from this already dialectic starting point, a development in which the pursuit of sense and the resistance of senselessness are considered in conjunction. In this dialectic development we will recognise the different forms of violence presented here, even though Weil does not refer to these forms in so many words.

In the first phase of this development, man overcomes the violence and resistance of nature through subjugation based on the counterforce of labour. At the same time this enables him to integrate into the wider realm to which he perceives himself to belong. Nature and the human project ideally exist in perfect harmony. Ripples in this relationship on an individual level—such as differences in harvest yields—are ironed out within the community. Violence has no place in this process, and anyone who rejects the communal harmony will be banished. The language of the community ('that is how we do things here') renders individual violence invisible. Perhaps one could say that this reveals one side of nationalist violence (although more accurately it is concealed rather than revealed), this being the inward facing side that unites the nation. The other side of that same violence manifests itself in the following phase, when this self-confident unity is broken.

The disruption of this communal harmony comes from outside, from the plurality of communities that emerge in such circumstances. In this second phase—the conflict situation—violence is aimed at those other communities, initially taking more or less the same form as the outwardly focused violence that previously targeted nature. At first glance the differences are negligible. Similar to the resistance offered by nature, the other community now manifests itself as a body of resistance against the social order of the own community with a contrasting identification of sense. The difference between the two situations is revealed in the reaction to violent attempts to break that resistance. While nature essentially remains passive and allows itself to be overwhelmed, the other community fights back, causing the violence to become increasingly blatant in conflicts between subjugators and subdued, and between oppressors and oppressed, taking on many different forms including revolution, emancipation and changing coalitions. This violence is expressed in the language of command and obedience. The violence described here takes place not just between nations but primarily between groups or classes within a nation, according to Eric Weil. In effect, in this phase the transition is made from the closed community of the first phase to a broader society. Society is therefore characterised by a dual battle: the external battle against nature and the internal battle between the classes. At the same time this clarifies the entanglement between these two forms of violence. The oppressors leave the battle against nature to be fought by the oppressed, while enjoying the fruits of this labour themselves.

This violence, whether in the form of a class war or a war between nations, is, as it were, eliminated in what could be referred to as the third phase, in which the violence of battles between people or groups is perceived to be counter-productive and needs to be eradicated through organisation: *'Les loups s'organisent donc entre eux'* (Weil, 1967, p. 81). In an organised society everyone is moulded into a useful cog in a comprehensive machine, which means on the one hand that everyone is valuable but on the other hand that everyone is objectified and commodified, thereby losing their individual identity. Eric Weil gives a shockingly recognisable depiction of the language in which what could be referred to as the structural violence of this organised world is expressed: serious language, the language of rational discussion, in which everything is objectified in order to enable compromise and objective agreement. It is the language of science, in which the scientist—the individual—is lost, and this is particularly evident in the field of human science. Individuals becomes objects that can be understood in terms of general, objective laws and no longer need to understand themselves. It leaves a particularly unpleasant aftertaste when Weil describes how this language initially accommodates the old language or languages, only to gradually replace them. People will gradually learn how to refer to themselves and their world in scientific terms: *'A la fin, tous parleront le langage de la rationalité et du calcul'* (p. 82). Until then old words such as justice, eminence, dignity, freedom and equality will be tolerated in the realisation that they have been liberated from their original, natural and social conditions. They may possibly have been sublimated, but Weil seems to suggest that they have also been severed from their original meaning, becoming void of meaning and nihilistic.

In the fourth phase, the greater the success of this language and this social organisation, the greater the failure. The more it succeeds in conquering the problems of nature and of the other, the more it leaves a world in which mankind itself becomes superfluous. The universalisation of rationality and calculation creates a situation of constant tedium from which it is only possible to escape through irrational violence, whereby the individual is defined in terms of opposition to others, in an affirmation of self through the negation of the other. This seems to be a form of senseless violence, of resistance to the language of rationality, which has become a lie in the extent to which it ignores sense.

What is exceptional about this development, or this way of describing the development, is that it is not only and not so much the violence itself that emerges as a constant danger, but rather the threat of suppression of violence. Not only do we see the continual elimination of violence, we also see how dangerous this is since it suppresses not just violence but also the desire to make sense of it. Violence is not comprehended or made comprehensible, is not made sense of (*'élevé ... au sens'),* but is denied and concealed. As stated earlier, Weil goes further than Ricoeur in examining how the opposition between language and violence also implies a connection: '*Seul l'être violent, s'il parle, peut chercher un sens*' (p. 85). Violence and sense are on opposing sides but are inseparable.

What does this all mean in practice? What should we do about violence? Weil does not advocate a return to nature, even if this were possible. On the contrary, mastering the savage violence of nature is historically one of mankind's greatest achievements. What we need to avoid is a situation in which we lose sight of the question of sense while learning to control the violence of senselessness. For this we need to acknowledge the relationship between violence and language. According to Weil, we will have to learn to perceive ourselves as a unit that incorporates both violence and language, a unit that has learned to distinguish between these two phenomena over the course of our development but cannot untangle them definitively or vanquish the tension between them definitively. To indicate how this should be done he refers to the Hegelian trinity of art, religion and philosophy, in which he believes that man simultaneously expresses, denies and transcends both the violence of the absurd and the violence of his own passionate nature.

Although I am not entirely sure what Weil means by this, I suspect that it is an important point. Human culture cannot exist in a world in which violence is eliminated, but must exist in a world in which it is sublimated and simultaneously serves as a reminder. In common with the external violence of the natural world, which cannot be completely eliminated since we have to stay alert to the possibility of floods, hurricanes and volcanic eruptions, the internal violence within mankind cannot be completely eliminated either. We will have to learn to live with the violence of disease and death since this cannot be banished even by perfecting our grasp of medical science; we will be unable to fully tame the violence of our desires without becoming a slave to our domestication; and we will be unable to fully eradicate the violence of senselessness using our designs of sense without becoming blind to the fragility of those designs. We will have to learn that violence is not outside of us or opposite us but rather that it comprises half of our being. It is not a question of replacing violence with language but of establishing an appropriate balance between the two. The problem of violence is therefore a question of degree.

References

Derrida, J. (1964). Violence et métaphysique: Essai sur la pensée d'Emmanuel Levinas. *Revue De Métaphysique Et De Morale, 69*(3), 322–354.

Levinas, E. (1987). *Collected philosophical papers*. Martinus Nijhoff Publishers.

Ricoeur, P. (1967). Violence et langage. In Collectif (Ed.), *Recherches et Débats, n°59: La Violence. Semaine des Intellectuels Catholiques* (pp. 86–94). Desclee de Brouwer.

Verhoeven, C. (1967). *Tegen het geweld [Against violence]*. Ambo.

Weil, É. (1967). Violence et langage. In Collectif (Ed.), *Recherches et Débats, n°59: La Violence. Semaine des Intellectuels Catholiques* (pp. 78–86). Desclee de Brouwer.

Paul van Tongeren (1950), studied theology and philosophy in Utrecht (Nl.) and Leuven (B.). Since 1993 and until his retirement in 2015, he was a full professor of moral philosophy and ethical theory in the Department of Philosophy of Radboud University in Nijmegen (NL). Since 2002 he was also an extraordinary professor of ethics at the Institute of Philosophy in Leuven (B.). His research focuses mainly on Nietzsche, on hermeneutics and on (the history of) moral philosophy. Currently, he is the Dutch national Thinker Laureate.

Military Trauma and the Conflicted Human Condition: Moral Injury as a Window into Violence, Human Nature and Military Ethics

Tine Molendijk

> 'This man on his motorcycle keeps ignoring my warning shots. Should I now take aim and shoot at the risk of killing an innocent man?' 'Our local ally oppresses the people living in the districts that he governs. Should I say something about it, even though I am not allowed to?' 'These people are hostile to our presence. Why are we even here?'

Questions like these testify to the complexities of soldiering in a conflict zone. The past decades have seen military operations become increasingly complex, also in an ethical sense. Contemporary missions are no longer solely interventions in interstate conflicts where two regular armed forces oppose and fight each other, but operations in what has been called irregular warfare. Today's soldiers are often confronted with internationalized intrastate conflicts and with unconventional fighters who use irregular tactics and are generally difficult to distinguish from civilians. Moreover, the roles and tasks of contemporary soldiers are often complex. They may have to fight, build relationships with local actors and develop humanitarian activities in one and the same mission (Baarda & Verweij, 2009; Molendijk, 2019). This complexity seems to be the reason why 'moral injury', which refers to the lasting psychological impact of morally critical situations such as tragic dilemmas and moral transgressions, has been called the 'signature wound' of contemporary missions (Pederson, 2021).

Different from post-traumatic stress disorder (PTSD), which is generally conceptualized as the result of a (life-)threatening event and is therefore predominantly associated with fear-related symptoms, moral injury is defined as psychological damage caused by '[p]erpetrating, failing to prevent, bearing witness to, or learning about acts that transgress deeply held moral beliefs and expectations' (Litz et al., 2009, p. 700). While moral injury symptoms can overlap with the symptoms associated with PTSD (e.g. nightmares and a negative mood), central to moral injury are profound feelings of guilt, shame and anger (Frankfurt & Frazier, 2016; B. J. Griffin et al., 2019).

The majority of current research on moral injury is psychological research focused on the psychometric properties, diagnosis and clinical treatment of moral injury

T. Molendijk (✉)
Department of Military Management Studies, Faculty of Military Sciences, Netherlands Defense Academy, Breda, The Netherlands

E.-H. Kramer and T. Molendijk (eds.), *Violence in Extreme Conditions*,
https://doi.org/10.1007/978-3-031-16119-3_3

(Frankfurt & Frazier, 2016; B. J. Griffin et al., 2019; Williamson et al., 2018). At the same time, a growing body of literature points out that the specific potential of the concept lies in drawing attention to the often complicated ethical and social dimensions of military suffering, aspects which have received marginal attention in trauma literature (Kinghorn, 2012; Molendijk, 2021; Scandlyn & Hautzinger, 2014; Shay, 2014). Indeed, when veterans speak about feeling guilty, ashamed and angry about what they have done, failed to do, or had to witness, they speak about an experience of moral conflict engendered by, for instance, confrontations with tragic dilemmas in which they had to choose between two evils (Baarda & Verweij, 2006; Molendijk, 2021; Rietveld, 2009; Sherman, 2015). Moreover, they often speak about moral disorientation and existential confusion, both within themselves and in relation to the organization, political leadership or society, by which they feel betrayed, alienated, or both (Lifton, 1973; Molendijk, 2021; Shay, 1994).

In this chapter, I consider moral injury in this sense; that is, in terms of moral conflict at the psychological micro level in relation to structural tensions at the level of the organization, politics and society. In doing so, I draw on Verweij's work on moral conflict as inherent to being human and as something that can manifest itself in destructive ways in extreme contexts. First, I explain moral conflict as inherently part of human nature. Next, I discuss how veterans' stories of moral injury offer important insights into violence, human nature and military ethics, at the level of the individual soldier's psyche, the relationship between soldier and society, and society at large. Finally, I discuss the implications of these insights for military ethics.

Moral Conflict as Inherent to Being Human

People are not made of one piece. As Verweij has convincingly argued throughout her work (Verweij, 1993, 2007, e.g. 2010a, 2010b; Verweij & Jespers, 2001), moral conflict is inherent to being human. These conflicts may include battles between good and evil, but more generally are struggles between constructive and destructive forces, where the destructive force is not even necessarily evil. The destructive force can be considered a necessary counterpart of the constructive force in the human psyche—and in society at large—which only produces evil when out of control. These are Verweij's words, as she traces back this insight to Ancient Greek philosophy:

> Pre-Socratic philosopher Empedocles explicitly used the words Love and Hate with reference to these powers; Plato used the term 'Eros' (Love) and in order to explain the power of Eros and its negation in the human soul, he used the concepts logos, epithymia and thumos, of which thumos turns into destructive energy when it is not balanced by Logos (Plato 1999). Freud (1991) and Nietzsche (1988) respectively referred to these two contraire [sic], yet mutually dependant [sic] powers as 'Eros' and 'Thanatos' and the 'Apollonian' and the 'Dionysian'. Hate, as destructive energy, can thus be located in Plato's thumos, as well as in Freud's Thanatos and Nietzsche's Dionysian, all on both an individual and a political level. What the concepts thumos, the Dionysian, and Thanatos have in common is a penchant for violence and destruction when this inner drive or energy is not counterbalanced. In that sense, Plato as well as Nietzsche and Freud underline the importance of the inherent

> coherence between the opposing forces that form the basis of human development both individually and collectively. (Verweij, 2018, p. 209)

This view of the human condition suffuses Verweij's approach to military practice and military ethics. War itself is 'violent and destructive' while it can also be 'justifiable and even morally necessary', she states, and human beings 'hurt and help other people and sometimes (…) do both things at the same time' (Verweij, 2007, pp. 44, 58). Thus, mankind is 'deinon' or dissonant, and while under normal, 'civilized' circumstances we are able to regulate our constructive and destructive forces, in war these latent forces may become manifest and produce evil. This is where the relevance of military ethics lies, according to Verweij: 'we need ethics in order to deal with our "deinon" character' (Verweij, 2007, p. 58).

Following this path, in this chapter I will consider psychological trauma, and moral trauma—'moral injury'—in particular, in relation to the dissonant character of human nature and the complexities following from this. War and violence can reveal and unleash dark forces in and between people. In turn, moral injury does something similar. Through psychiatric symptoms, the phenomenon of moral injury sheds light on tensions in and between people, showing how tensions that are otherwise manageable may in some circumstances (such as war and violence) grow into unbearable pressures causing a mental breakdown. To draw on Freud's crystal metaphor, '[i]f a crystal is thrown to the ground, it will break into pieces, not in a random way, but according to specific fault-lines which, although they are invisible, have been predetermined by the structure of the crystal' (Freud, cited in Corveleyn, 2009, p. 87). Among other things, the fault lines that moral injury makes manifest are people's moral beliefs and friction between them, as well as friction between people's moral beliefs and the social worlds in which they are embedded. Moreover, moral injury offers insight into tensions existing in society at large.

Moral Injury as a Window into Violence, Human Nature and Military Ethics

I have identified three categories of moral tension, playing out on three levels, that manifest themselves in moral injury. Put differently, I found that veterans' stories of moral injury offer three important insights into violence, human nature and military ethics. First, morality's complexity and the dissonant human condition; second, the friction-inducing embeddedness of soldiers' experiences in various social contexts; and third, the uncomfortable tension existing in society at large with regard to military practice.

Tension as Part of Morality and the Human Condition

The complexity of our moral beliefs and our condition as humans in general was painfully articulated by philosopher and World War II veteran Jesse Glenn Gray in his celebrated book *The Warriors: Reflections on Men in Battle* (Gray, 1959). The book's longest chapter, which is based on his experience as a soldier in what we think of as the 'Good War', is called *The Ache of Guilt*. Here he quotes his own wartime diary, describing several deaths for which he was directly or indirectly responsible, concluding: 'I hope it will not rest too hard on my conscience, and yet if it does not I shall be disturbed also' (Gray, 1959, p. 176).

Reflecting on what war brought out in him and his fellow soldiers, he writes:

> The reflective soldier on both sides of the conflict will see no escape from political guilt as long as he remains a member of a state. If, in his disillusionment, he is tempted to renounce his nation and pledge his allegiance to the human race alone, this, too, will prove illusory, for mankind collectively is doubtless as predisposed to injustice as nations are. (…) Faced with this presumptuousness of the human creature, his closedness and dearth of love, the awakened soldier will be driven to say in his heart: 'I, too, belong to this species. I am ashamed not only of my own deeds, not only of my nation's deeds, but of human deeds as well. I am ashamed to be a man.' (…) How many soldiers have experienced in battle a profound distaste for the human creature! (Gray, 1959, pp. 205–207)

The moral turmoil Glenn Gray describes here is existential. As he writes, the violence he witnessed and participated in engendered 'a guilt that can only be called metaphysical, because it concerns man's very being and its relations to the rest of the cosmos' (Gray, 1959, p. 206).

Glenn Gray's struggle testifies to something more complex than a realization that human beings are simply evil. The fundamental issue 'lies in human nature itself, in our failure as human beings to live in accordance with our potentialities and our vision of the good' (Gray, 1959, p. 206). The painful realization is that the notions of good and evil in the way they are usually understood are untenable. It is the realization that the world of war, the world of so-called 'inhumanity', is not a world distinct from the human universe but the very opposite. Only human beings are able to be inhumane—we would not call a lion killing an African buffalo calf such—and this is exactly why exposure to inhumanities, including people's own inhumanity, can wound them so profoundly.

The conflicts that define being human are therefore not so much about good versus evil—if only it were so clear cut. They are about constructive versus destructive forces. Moreover, the destructive forces are not necessarily evil. For instance, forces of indignation may also fuel a passion to courageously fight for what one believes is right. This is what Plato meant by *thumos*: the third part of the soul by virtue of which we feel righteous anger, the desire to combat perceived injustice and the duty to uphold our honour (Verweij & Jespers, 2001). To reiterate, the destructive forces may result in dark destruction only when they are out of control.

In a similar vein, a person's moral beliefs are not a clear systematic framework of morals that as long as rigidly followed protect the person against immoral behaviour. Moral beliefs are a 'messy', time and place-dependent constellation of values that may conflict with other drives and with one another (Molendijk, 2021). This is not just a theoretical issue. There are practical implications. The critical situations that soldiers may face during deployment are usually not unambiguous moral transgressions, but confusing dilemmas where in order to respect one value they are forced to violate another, and where being a good soldier may feel like being a bad person. Also, soldiers may engage in acts of moral disengagement, where they transgress moral boundaries that they would not transgress under normal circumstances, because these boundaries are less clear cut at that moment. These types of experiences are more complex than an unambiguous feeling that the soldier has *undeniably* violated his or her moral code, or that soldiers undeniably deviated from their moral compass. Such experiences involve conflict (Molendijk et al., 2018; Verweij & Molendijk, 2019).

The Frictional Embeddedness of Soldiers in Various Social Contexts

The experience of inner conflict is related to a second insight that stories of moral injury reveal concerning friction between soldiers and the various social contexts in which they are inextricably embedded. Philosopher and Vietnam veteran Camillo Mac Bica experienced this friction first hand on returning home from war, which he came to call a 'moral identity confusion'. The confusion he describes involves a profound sense in soldiers of being in limbo between two worlds, 'the world they recognized as their place of origin – though, now, quite foreign and inhabited by alien though recognizable individuals they had once loved – and the world of killing and destruction – of which they now feel a part' (Bica, 1999, p. 89).

Some of the Dutch veterans I spoke with for my own research on moral injury described similar feelings, and ones that were even more complicated. Most had not turned against war and the military. Indeed, as they told me once a rapport had been established, many felt no shame whatsoever for having fought and killed opponents, but struggled with the fact that 'people at home' felt they should. Many Afghanistan veterans, for instance, told me that they got 'a kick' out of engaging in combat and often felt ˈhomesickˊ for their deployments when thinking about it.

This does not mean that the positive feelings soldiers may describe can simply be put down to 'fun'. Rather, the veterans I spoke with used words such as 'good' and 'unique'. They often described a confluence of antagonistic feelings, including fear, the effects of adrenaline and excitement, which is reminiscent of how Vietnam veteran and novelist Tim O'Brien describes war stories. As he writes:

> war is mystery and terror and adventure and courage and discovery and holiness and pity and despair and longing and love. War is nasty; war is fun. War is thrilling; war is drudgery. War makes you a man; war makes you dead. (O'Brien, 1990, pp. 86–87)

In moral terms, this means, as O'Brien writes, that a 'true war story does not instruct, nor encourage virtue, nor suggest models of proper human behaviour. (…) Send guys to war, they come home talking dirty' (O'Brien, 1990, p. 76).

Yet, this is not something that 'people at home like to hear', as many veterans told me. Take what one former private told me about his deployment to Afghanistan. He commented that at the time, especially after he had lost a close colleague, he 'really wished' that he had been able to see the guys he had possibly 'whacked' up close. He never really spoke about his experience with civilians. When I asked why, he sketched the following scenario, in which 'I' is a civilian and 'this guy' is a military veteran.

> Okay, so imagine, I don't know this guy and suddenly I ask, 'Did you ever kill someone?' And he says 'Yes', without blinking. Later, I go to a mate of mine. 'Listen to this', I say. 'There's a guy and he says without blinking that he's killed someone. He's fucking sick in the head man.' It's just a nice story, so people have something to tell each other, they don't really care about it in any way other than that.

Negative experiences like these led many soldiers to stop telling people that they were in the military.

The soldiers I spoke with often tied people's perceptions of military practice to societal double standards regarding the military. One was a soldier who had served in Afghanistan as section commander. Although he had been against the political motives for sending troops to Afghanistan, he had been in favour of the mission. But back home, he said, 'You have to come up with an excuse for why you are in the military'. According to him, many Dutch civilians like to see violence 'as something sad and horrible', because it is 'nice and easy' to see it that way, while also enjoying the privilege of having armed forces. He elaborated on this point as follows:

> We don't want to know about the price we pay for it. It's like we want to eat meat, but don't want to know how the cows were butchered. (…) We only want to eat meat that was obtained without violence, but that's of course impossible, to get meat without violence.

Note that he said 'we' rather than 'they don't want to know'. When I mentioned this to him, he replied it was not a mistake, and that 'I also know that, if we'd really follow our moral compass, we'd be in Africa or somewhere else right now, where the really serious conflicts and genocides are, but that's not where our interests lie'. Many veterans, while lamenting civilians' lack of understanding, also admitted having mixed feelings about their profession. They experienced a duality in civilians, in themselves, and in themselves vis-à-vis civilians.

The more general insight offered by stories like these is that the duality within people also manifests itself in the relationship between soldiers and society. Moreover, the stories reveal that these two dualities should not be understood in binary terms, as the sum of inner tension within people plus social tension between people and their social environment. People are *always* socially embedded, from birth, and their inner lives are thus always socially embedded and shaped as well. People live and act on a daily basis within a range of social levels (e.g. group, organization, nation) and a range of social contexts (e.g. different ethnic cultures and professional cultures). In doing so, they develop assumptions and meanings through which they understand their experience and make judgements about what is acceptable and unacceptable conduct, creating a moral compass that guides their actions (Bandura, 1991; Haidt & Joseph, 2004; Hitlin & Vaisey, 2013). Yet, as the world is not a coherent, harmonious system, this moral compass is not a simple one-needle device guiding the way but a messy constellation of commitments that do not always align.

It is therefore not only because people have dual drives, but also because soldiers are embedded in various social worlds that they embody multiple commitments that at times make conflicting demands, thereby creating tensions that they need to manage. When these tensions turn out to be unmanageable, the result may be a profound existential crisis, both within soldiers and between them and the people around them. They may feel deeply disillusioned, disoriented and alienated, from both themselves and the world (Molendijk, 2021).

Tensions in Audiences of Stories of War and Moral Injury

A final field of tension coming to the surface in stories of moral injury lies in the audiences of these stories. As the duality within people also manifests itself in society at large, this has implications not only for how soldiers relate to civilians, but also for how civilian audiences, each with their own values and assumptions, hear and retell soldiers´ stories.

Again, let me offer an illustration. In 2020, Dutch newspaper *Trouw* published an interview with a Dutch veteran about an incident he experienced in Afghanistan in 2007. The veteran recounted that he had shot at civilians at the order of his commander and killed them, while in hindsight he was sure that they had not shown any hostile behaviour (Ziel, 2020). The incident had never stopped haunting the veteran, and as he felt the military would not take his story seriously, he decided to confess in public. The Netherlands Ministry of Defence responded by asserting that no After Action Report or other document could be found in the archives about the incident described by the veteran. However, an investigation was launched, which is still ongoing at the time of writing.

I know the veteran through my research. While his confession of a possible war crime was new, he had spoken before about civilian casualties. After returning home from his deployment, he had developed a great sense of sadness, which had soon changed into anger and paranoia. He had suspected his loved ones of trying to do something to him. After eventually seeking help, he was diagnosed with PTSD and was given a psychiatric assistance dog, which now wakes him up before he wakes himself up with his own screams, dreaming of the events in Afghanistan. In his nightmares, dead Afghan civilians still approach him, asking, 'Why did this happen?'.

The weeks following the publication of the interview in the newspaper, both civilians and veterans reached out to me, because in the interview the veteran in question had mentioned that he might be suffering from 'moral injury' (which they knew was a topic I focused on). Some people, mostly soldiers and veterans, angrily called the veteran a 'traitor'. They did not believe his story and thought that he had added fuel to the fire regarding the already negative image many Dutch civilians have of the Dutch military. Others, mostly civilians, expressed shock and dismay at the incident and sympathized with the veteran in question, sometimes adding statements along the lines of 'this is why we should not send people on immoral missions like these'. Yet others, mostly soldiers and veterans again, lamented such responses from civilians as wrong and detrimental to veterans, but also stated that 'this is why we should appreciate our veterans more and give traumatized buddies the psychological help they need'. On social media, I read many messages similar to those I received.

These kinds of responses can be observed in relation to the concept of moral injury more generally. The concept has been embraced by both supporters and opponents of the military, and usurped by some to direct attention towards a broader ideological issue. For instance, I have observed advocates for veterans, researchers and professionals put forward moral injury as an indication that veterans need to be appreciated more and society needs to have a more positive image of what soldiers do (see also Morris, 2015). Conversely, others interpret moral injury as the result of war's inherent immorality (Meagher, 2014) and, in this light, see the morally injured veteran as a 'prophet' (Antal et al., 2019) who can see 'deeply and radically into the truth of the present and how one's country is actually affecting others throughout the world' (Wiinikka-Lydon, 2017, p. 228). Importantly, these two interpretations have something in common. In both cases, veterans' stories of moral injury are retold as testimonies of wisdom about wider processes in politics and society, in ways that tend to say more about the beliefs of the advocates, researchers and professionals retelling the stories than of the beliefs of the morally injured veterans.

Indeed, research among both Dutch and US veterans indicates that many veterans actually feel alienated by societal pro-veteran praise, especially when they struggle with what they have done or have not been able to prevent (Bica, 1999; Lifton, 1973; Molendijk, 2021). Veterans told me, for instance, that they felt 'weird' and 'extremely guilty' when people spoke about them as heroes. At the same time, as stated, only a minority become anti-war. Their criticism is often directed towards the particular mandate and rules of engagement of a mission, not towards the mission or military intervention in itself (Drescher et al., 2013; Molendijk, 2021). So, without wanting to invalidate the above-mentioned ideological claims, they can have counterproductive consequences, because moral injury is understood in ways that are not so much directed towards the needs of the veterans as towards the views of the advocates, researchers and professionals voicing these claims.

In any case, these responses reveal that the concept of moral injury not only refers to morally significant experiences, but has great moral significance *as a concept*, carrying, and being charged with, normative claims about the nature and causes of moral conflict-coloured suffering. More broadly speaking, these responses show that not just veterans struggle with the moral significance of military intervention, but society does as well. Perhaps part of this discomfort lies in that it reveals a glimpse of the truth that we all have destructive drives within us (cf. Pellón, 1988). Whatever the reason, military intervention and the violence that is inevitably part of it causes discomfort in all of us.

Historically, this discomfort has been expelled by moving violence to the margins of society. Violence has been outsourced to the armed forces, which have to operate out of sight of society while, at the same time, their actions are closely monitored with a degree of suspicion (Bredow, 2006; De Swaan, 2015; Pellón, 1988). In terms of representation, too, the destructive dimension of military practice is carefully kept out of the frame (M. Griffin, 2010). What is in the frame is either 'sanitized' with romantic, heroic imagery in order to accept it back into the realm of the normal and justifiable or condemned as immoral, whereby the destructive dimension of military practice is either completely externalized to the soldiers, who are made the perpetrators doing the dirty work, or internalized to soldiers racked with guilt as a sinfulness that we all share (cf. Bredow, 2006; Pellón, 1988; Whitehead, 2004). The pro-veteran and anti-war interpretations of moral injury discussed above can also be understood as such. By reducing the complicated, dual, ambivalent experiences of veterans to binaries of 'wrong' and 'right', both interpretations resolve any tension related to military intervention.

Implications for Military Ethics

Thus, at the level of the soldiers' psyche, the soldier in relation to society and society at large, some fundamental moral tensions exist which are not easily resolved. This has particular implications for the question of how to approach military ethics. Verweij's work is again insightful here. According to her, the focus of military ethics should extend beyond rules and regulations, which 'only form a necessary, albeit insufficient, condition for moral competence' (Verweij, 2007, p. 59). Given the conflicts that are inherent to human nature and morality, approaching military ethics solely in terms of rule-following and impeccable conduct is unrealistic and even undesirable. Such an approach, moreover, can easily become a form of window dressing in the shape of what Nietzsche (an important inspirational source for Verweij) called *Aushange-Tugenden*, or signboard values (Verweij, 2010b).

Moral competence, as Verweij therefore argues, is not so much about abiding by ethical rules and codes of conduct. Rather, it is about being able to identify and deal responsibly with conflicting values in, for instance, moral dilemmas (Baarda & Verweij, 2006). It is, in line with how Cox and Calhoun conceptualize integrity, not about exhibiting legally and morally impeccable conduct but, rather, about 'the ability to constructively deal with conflicts within yourself and with others' (Verweij, 2010b, p. 16, translation TM). It is 'an ability based on critical reflection, which also implies critical self-reflection' and 'the ability to recognize and constructively deal with the doubts, conflicts and inner tension associated with this reflection' (Verweij, 2010b, p. 20, translation TM).

Besides acknowledging the dual human condition and the paradoxes of military practice, military ethics should go beyond the individual level (Verweij & Molendijk, 2019). If only dealt with by training and educating soldiers in ethics in order to strengthen their moral competence, military ethics loses sight of all the other factors and actors involved in the development of morally critical situations, if this is the case, military ethics may even contribute to feelings of alienation and betrayal among soldiers confronted with morally critical situations rather than help them in dealing with such situations (Molendijk, 2021).

It would be both appropriate and helpful to acknowledge that morally responsible military intervention is only partly within soldiers' own control. Military practice is a collective affair. The missions on which military personnel are to be deployed, and what they should and should not do in the area of operations, are determined at the political level, and debates about whether or not a mission was justified and useful are held at a wider social level. This is not just an issue that must be acknowledged with respect to soldiers. It must also be acknowledged at organizational, political and societal levels. Take ethical principles such as those of the just war tradition, which (despite what the name suggests) acknowledge rather than obscure that 'war' and 'justice' exist in fundamental tension with one another. These ethical principles are already embedded in, for instance, international humanitarian law (including the Geneva Conventions and the Charter of the United Nations), but in practice they are often ticked off as if they are part of a checklist rather than genuinely taken into account as ethical principles (Verweij & Molendijk, 2019). Such a checklist approach renders these principles almost meaningless, as it brushes over their complexities, not to mention that it looks more like 'sanitizing' window dressing than ethical decision-making. Instead, a sincere, careful consideration of the purposely complicated and even paradoxical notion of 'just war' may help decision-makers and society in general acknowledge and manage the tensions and discomfort of military practice.

This brings me to a final point, namely the issue of us hearing and retelling stories of moral injury in ways that actually distort these stories and say more about us than the affected individuals who told us the stories in the first place. As a society, we are simultaneously fascinated and uncomfortable when it comes to military intervention, and we tend to resolve this discomfort with unequivocal interpretations of such an event. Therefore, as researchers, professionals and others interested in moral injury, we should be particularly aware of potential tendencies in ourselves to impose our own moral beliefs on the stories we hear and to readily approach tensions and conflicts as 'kinks' that need to be ironed out. When acknowledging tension and discomfort in others and ourselves, it becomes possible to illuminate aspects of soldiers' experience that might otherwise be obscured.

Conclusion

This chapter argues that stories of moral injuries—and trauma more generally—not only shed light on the injuries themselves, but also bring to the surface several broader tensions and vulnerabilities underlying these injuries, which normally remain hidden. To again use Freud's crystal metaphor, this chapter's investigation of the crystal pieces of moral injury offers insight into the fault lines of the human psyche and into organizational and sociopolitical fault lines. It makes manifest moral tensions that military practice and its contextual aspects generate in soldiers in general, whether or not they develop psychological distress as a result of these tensions, and reveals basic vulnerabilities of which it is vital that they are addressed in military ethics.

References

Antal, C. J., Yeomans, P. D., East, R., Hickey, D. W., Kalkstein, S., Brown, K. M., & Kaminstein, D. S. (2019). Transforming veteran identity through community engagement: A Chaplain-psychologist collaboration to address moral injury. *Journal of Humanistic Psychology*, 1–26.

van Baarda, T. A., & Verweij, D. E. M. (2006). *Military ethics: The Dutch approach - A practical guide*. Martinus Nijhoff.

van Baarda, T. A., & Verweij, D. E. M. (2009). *The moral dimension of asymmetrical warfare: Counter-terrorism, democratic values and military ethics*. Martinus Nijhoff.

Bandura, A. (1991). Social cognitive theory of moral thought and action. In W. M. Kurtines & J. L. Gewirtz (Eds.), *Handbook of moral behavior and development* (pp. 69–128). Erlbaum.

Bica, C. C. (1999). A therapeutic application of philosophy. The moral casualties of war: Understanding the experience. *International Journal of Applied Philosophy, 13*(1), 81–92.

von Bredow, W. (2006). The order of violence: Norms, rules, and taboos of organized violence and the de-legitimization of the military. In G. Caforio (Ed.), *Handbook of the sociology of the military* (pp. 87–98). Kluwer Academic/Plenum Publishers.

Corveleyn, J. (2009). Religious delusion in psychosis and hysteria. In J. Belzen A. (ed.), *Changing the scientific study of religion: Beyond freud? Theoretical, empirical and clinical studies from psychoanalytic perspectives* (pp. 85–109). Dordrecht: Springer. http://link.springer.com/content/pdf/https://doi.org/10.1007/978-90-481-2540-1.pdf#page=86. Accessed 24 October 2017.

De Swaan, A. (2015). *The killing compartments: The mentality of mass murder*. Yale University Press.

Drescher, K. D., Nieuwsma, J. A., & Swales, P. J. (2013). Morality and moral injury: Insights from theology and health science. *Reflective Practice: Formation and Supervision in Ministry, 33*(1), 50–61.

Frankfurt, S., & Frazier, P. (2016). A review of research on moral injury in Combat Veterans. *Military Psychology, 28*(5), 318–330.

Gray, J. G. (1959). *The warriors: Reflections on men in battle*. University of Nebraska Press.

Griffin, B. J., Purcell, N., Burkman, K., Litz, B. T., Bryan, C. J., Schmitz, M., et al. (2019). Moral injury: An integrative review. *Journal of Traumatic Stress, 32*(3), 350–362.

Griffin, M. (2010). Media images of war. *Media, War & Conflict, 3*(1), 7–41.

Haidt, J., & Joseph, C. (2004). Intuitive ethics: How innately prepared intuitions generate culturally variable virtues. *Daedalus, 133*(4), 55–66.

Hitlin, S., & Vaisey, S. (2013). The new sociology of morality. *Annual Review of Sociology, 39*(1), 51–68.

Kinghorn, W. (2012). Combat Trauma and moral fragmentation: A theological account of moral injury. *Journal of the Society of Christian Ethics, 32*(2), 57–74.

Lifton, R. J. (1973). *Home from the war: Learning from Vietnam veterans*. Other Press.

Litz, B. T., Stein, N., Delaney, E., Lebowitz, L., Nash, W. P., Silva, C., & Maguen, S. (2009). Moral injury and moral repair in war veterans: A preliminary model and intervention strategy. *Clinical Psychology Review, 29*(8), 695–706.

Meagher, R. E. (2014). *Killing from the inside out: Moral injury and just war*. Cascade Books.

Molendijk, T. (2019). The role of political practices in moral injury: A study of Afghanistan Veterans. *Political Psychology, 40*(2), 261–275.

Molendijk, T. (2021). *Moral injury and soldiers in conflict: Political practices and public perceptions*. Routledge.

Molendijk, T., Kramer, E.-H., & Verweij, D. E. M. (2018). Moral aspects of 'moral injury': Analyzing conceptualizations on the role of morality in military trauma. *Journal of Military Ethics, 17*(1), 36–53.
Morris, P. J. (2015). Caring for heroes. *North Carolina Medical Journal, 76*(5), 293–293. https://doi.org/10.18043/ncm.76.5.293.
O'Brien, T. (1990). *The things they carried*. Houghton Mifflin Press.
Pederson, J. (2021). 'Signature wound': moral injury in Iraq war literature. In *Sin Sick: Moral injury in war and literature* (pp. 125–155). Cornell University Press.
Pellón, G. (1988). Myth, tragedy and the scapegoat Ritual in Crónica de una muerte anunciada. *Revista Canadiense De Estudios Hispánicos, 12*(3), 397–413.
Rietveld, N. (2009). *De gewetensvolle veteraan [The Conscientious Veteran]*. Doctoral dissertation. Tilburg: Tilburg University.
Scandlyn, J., & Hautzinger, S. (2014). Collective reckoning with the post 9/11 wars on a Colorado Homefront. In *Costs of War*. Providence, RI: Watson Institute for International Studies.
Shay, J. (1994). *Achilles in Vietnam: Combat trauma and the undoing of character*. Simon and Schuster.
Shay, J. (2014). Moral injury. *Psychoanalytic Psychology, 31*(2), 182–191.
Sherman, N. (2015). *Afterwar: Healing the moral wounds of our soldiers*. Oxford University Press.
Verweij, D. E. M. (1993). *Ariadne en Dionysos: Vrouw-metaforen en verlangen in het werk van Nietzsche* [*Ariadne and Dionysos: Woman Metaphors and Desire in the Work of Nietzsche*]. Amsterdam: Doctoral dissertation. Radboud University.
Verweij, D. E. M. (2007). Military ethics: A contradiction in terms? In J. Toiskallio (Ed.), *Ethical education in the military: What, how and why in the 21st century?* (pp. 43–62). National Defense University.
Verweij, D. E. M. (2010a). *Geweten onder schot: Ethiek en de militaire praktijk*. Boom.
Verweij, D. E. M. (2010b). Integriteit of slechts een opgepoetst blazoen? In S. Verheij & E. Karssing (eds.), *Jaarboek Integriteit 2011* (pp. 14–24). Den Haag: Bureau Integriteitbevordering Openbare Sector (BIOS).
Verweij, D. E. M. (2018). The hate trail: Counter-terrorism of the 21st century and its challenges. In M. Badde-Revue & M.-D.-N. Ruffo de Calabre (eds.), *Ethics in counter-terrorism* (pp. 207–224). Brill Nijhoff. Accessed 10 July 2021.
Verweij, D. E. M., & Jespers, F. P. M. (2001). *Passie en persoonlijkheid: de thematiek van het verlangen belicht vanuit de filosofie en de psychopathologie*. Uitgeverij Van Gorcum.
Verweij, D. E. M., & Molendijk, T. (2019). Voorkomen is beter dan genezen: Moral injury vanuit macro- en micro-perspectief [Prevention is better than cure. Moral injury from a micro and macro perspective]. In M. van der Giessen, P. H. Kamphuis, E. R. Muller, U. Rosenthal, G. Valk, & E. Vermetten (eds.), *Veteranen: Veteranen en veteranenbeleid in Nederland* [*Veterans: Veterans and Veteran Policy in the Netherlands*] (pp. 165–184). Deventer: Wolters Kluwer.
Whitehead, N. L. (2004). On the poetics of violence. In N. L. Whitehead (Ed.), *Violence* (pp. 55–77). School of American Research Press.
Wiinikka-Lydon, J. (2017). Moral injury as inherent political critique: The prophetic possibilities of a new term. *Political Theology, 18*(3), 219–232.
Williamson, V., Stevelink, S. A. M., & Greenberg, N. (2018). Occupational moral injury and mental health: Systematic review and meta-analysis. *The British Journal of Psychiatry, 212*(6), 339–346.
Ziel, V. der Z. (2020, December 23). Militair servie Hölzken ging volledig over de schreef in Uruzgan. 'Het moet verteld worden'. [Soldier Servie Hölzken completely crossed the line in Uruzgan. 'It must be told']. *Trouw*.

Tine Molendijk is an interdisciplinary-minded cultural anthropologist specialized in the topics of violence, military culture and mental health, in particular post-traumatic stress disorder (PTSD) and moral injury. Her PhD research focused on the role of political practices and public perceptions in moral injury in Bosnia and Afghanistan veterans. Currently, she works as Assistant

Professor at the Faculty of Military Sciences at the Netherlands Defence Academy (NLDA) and as a Research Fellow affiliated to the Radboud University. Among her publications is the award-winning book Moral Injury and Soldiers in Conflict: Political Practices and Public Perceptions (Routledge, 2021).

Exploring the Relevance of the Systems Psychodynamic Approach to Military Organizations

Eric-Hans Kramer, Max Visser, and Matthijs Moorkamp

The psychological demands of the military profession have been a focal point of reflection and theorizing in the realm of military studies for a long time (Shephard, 2001). The extreme conditions in which military organizations operate can expose soldiers to violent behaviour and may require them to use violent force. Furthermore, they may encounter human suffering of various kinds and may even witness abuses of power. Such experiences may lead to moral conflicts in soldiers (Lifton, 1973; Molendijk, 2021), to the experience of losing one's existential foundation (Bica, 1999) and to psychological trauma (Grossman, 2009; Shay, 1994). While this subject has attracted the attention of (clinical) psychologists and, to a lesser extent, philosophers, sociologists, anthropologists and historians, it has rarely been approached from the perspective of organizational science. Considerable attention has indeed been paid to how culture and leadership influence individual ways of coping with the psychological demands of the military profession (Bica, 1999; Shay, 1994). However, the relationship between structural features of military organizations and the means available to operators for dealing with the psychological demands intrinsic to their profession has remained an underdeveloped area.

This theme is the focus of the systems psychodynamic perspective, which is a tradition in organizational science that relates the structural features of organizations to the psychodynamic aspects of the functioning of groups and individuals,

E.-H. Kramer (✉)
Department of Military Management Studies, Faculty of Military Sciences, Netherlands Defense Academy, Breda, The Netherlands
e-mail: F.J.Kramer@mindef.nl

M. Visser · M. Moorkamp
Institute for Management Research, Nijmegen School of Management, Radboud University, Nijmegen, The Netherlands
e-mail: max.visser@ru.nl

M. Moorkamp
e-mail: matthijs.moorkamp@ru.nl

E.-H. Kramer and T. Molendijk (eds.), *Violence in Extreme Conditions*,
https://doi.org/10.1007/978-3-031-16119-3_4

pointing at an interaction between the two (Gould, 2001; Gould et al., 2001; Krantz, 2010; Petriglieri & Petriglieri, 2020; Miller & Rice, 1967). The phrase 'structural features' is used here in a broad sense to refer to organization design, task design, work schedules and standard operating procedures; that is, to formal features of work systems that are designed to achieve a functional purpose. The core idea of the system psychodynamic perspective is that an organization is not just a formal work system but also a psychological environment (Petriglieri & Petriglieri, 2020), and characteristics from both realms mutually influence each other. According to the systems psychodynamic perspective, effective work systems should provide protection against the psychological demands that are intrinsic to a profession. However, the relationship between the two can become counterproductive, not just from the perspective of the psychological health of individuals but also from the perspective of the organization as an effective work system.

While this may seem straightforward, there are two main distinctions between the systems psychodynamic approach and more conventional approaches to work stress. First, the systems psychodynamic approach developed out of the psychodynamic approach in psychology and focuses on the problem of anxiety. Second, the emphasis is on the mutual relations between structural features and psychological demands. This means that the focus is on the ways in which structural features influence the means available for coping with psychological demands and vice versa. A main insight of the systems psychodynamic approach is that certain strategies for dealing with anxiety can become institutionalized in structural features—engraved in stone, as it were. In such cases, counterproductive defences against anxiety can become systemic and a foundation of professional socialization.

This chapter explores the relevance of the central insights of the systems psychodynamic perspective for the military organization. Because of the potential encounters with violence, the military organization is a prototypical example of an organization in which the core of the profession itself potentially generates anxiety. The problem of dealing with anxiety may therefore leave its traces in particular features of different military organizations. The exploration in this chapter focuses on two main issues. We discuss historical examples to reflect on the differences between armed forces that sustained both a form of social psychological integration and operational effectiveness and those that did not. Subsequently, we focus on contemporary missions and take a look at how their characteristic structural configurations influence the ability of operators to deal with the psychological demands of such missions. We start, however, by discussing the core of the systems psychodynamic perspective.

The Systems Psychodynamic Perspective

The systems psychodynamic perspective originated from the Tavistock Institute in London in the 1950s and 1960s. Initial work on the perspective developed from Bion's work on group dynamics, which applied psychoanalytical concepts to organizational contexts (Petriglieri & Petriglieri, 2020). Ultimately, this culminated in the book

Systems of Organization (Miller & Rice, 1967). Gould (2001, pp. 2–3) points out that the essence of this perspective is expressed in the conjunction of the terms 'systems' and 'psychodynamic'. On the one hand, concepts from open systems theory are used for understanding the structural aspects of organizations (design, divisions of labour, hierarchical relations, etc.) and the challenges organizations face in complex environments. On the other hand, psychoanalytic concepts are used (such as the unconscious, resistance, denial and regression). The conjunction of the two domains results in the study of the interaction between collective structures, practices, norms, and motivations and emotions in organizations. The starting point for the development of this approach was the idea that certain professions may, because of external threats or internal conflicts, intrinsically ignite anxiety that can manifest itself in disturbing affects and emotions.

Menzies Lyth (1988, p. 78) proposed the hypothesis that 'the success and viability of a social institution are intimately connected with the techniques it uses to contain anxiety'. Her paradigmatic study is that of nurses in a training hospital. Intrinsically, the nursing profession generates a complex array of anxieties, as a result of which nurses are at risk of becoming flooded by intense and unmanageable feelings (1988, p. 50):

> Nurses are confronted with the threat and reality of suffering and death as few lay people are. The work situation involves carrying out tasks which, by ordinary standards, are distasteful, disgusting, and frightening. [...] The work situation arouses very strong feelings in the nurses: pity, compassion, and love; guilt and anxiety; hatred and resentment of the patients who have aroused these strong feelings; envy of the care given to patients. (Menzies Lyth, 1988, p. 48)

Since these feelings intrinsically originate from the profession itself, they are referred to by Menzies Lyth as *primary anxieties*. Menzies Lyth's main and innovative point was that protections against such anxieties can become institutionalized in 'social defences against anxiety'. These are collective arrangements that protect members of an organization from disturbing affects (Petriglieri & Petriglieri, 2020). What makes this viewpoint innovative is that structural arrangements in organizations are not only viewed from a functional perspective in which psychological effects are an accidental by-product. Instead, psychological demands of the work are seen as a potentially important force that can explain the way that the structural arrangement of organizations are shaped. One example in Menzies Lyth's hospital is that of splitting up the nurse-patient relationship by means of the roster. The latter was constructed in such a way that it restricted the contact between individual nurses and patients, which prevented those nurses from being excessively confronted with situations that provoke primary anxiety (1988, pp. 51–53).

While protection against primary anxiety is important, the concept of 'social defences against anxiety' carries a negative connotation. It refers to mechanisms that help to solve primary anxieties in a primitive, ineffective and perhaps even counter-productive way. Their essence is that they eliminate the experience of anxiety, guilt, doubt and uncertainty (Menzies Lyth, 1988, p. 63) without helping individual nurses to handle such emotions constructively. They prevent professionals from experiencing 'the satisfaction and lessening of anxiety that come from knowing they have the ability to carry out their work realistically and effectively' (Menzies Lyth, 1988, p. 65). Other examples of social defence mechanisms were avoidance of change and an obscurity in the formal distribution of responsibility. Social defences against anxiety might even become embedded in certain traditional professional values and symbols, such as uniforms, which connote nurses as interchangeable agglomerations of nursing skills without individuality (Menzies Lyth, 1988, p. 52). Ineffective defence mechanisms might create other problems. The highly prescriptive and rigidly defined tasks in Menzies Lyth's hospital caused operational inflexibility, which in turn triggered the *secondary anxiety* in nurses of not being able to cope with everyday problems. The characteristics of the working environment resembled that of the prototypical rigid bureaucracy: minutely prescriptive rules and working practices and few opportunities for mutually supportive team relationships. In other words, the training hospital was a professional environment that structurally induced ineffective ways of dealing with primary anxieties, which subsequently created secondary anxieties that led to stresses and dissatisfaction (Menzies Lyth, 1988, p. 65).

Given that in certain organizations anxieties are intrinsic to the work itself, the question is how to protect professionals in a constructive way. Of importance in this regard is the development of a social context that reduces disturbing affects and facilitates sensemaking (Petriglieri & Petriglieri, 2020, p. 426). One way to create such an environment is by organizing social support for professionals. The concept of *sentient system* is used to refer to the combined arrangements within organizations that are meant to satisfy the emotional needs of members (Miller & Rice, 1967). The core idea is that task systems and sentient systems should overlap in such a way that psychological support matches the demands of a task, but as Petriglieri and Petriglieri (2020, p. 422) point out, 'Perfect overlap, however, is rare. Conflicts often arise between task and sentient systems, especially in times of change'. When task and sentient systems overlap, an organization has a developed structural arrangement that both supports effective problem solving and provides adequate support for professionals facing intrinsic anxieties.

Historical Dimension

Historically, armies have differed in the degrees to which their task systems and sentient systems overlapped. A comparison between the German and the US Armies of World War II seems most instructive in this regard (Dupuy, 1985; Hart, 2001; Van Creveld, 1983; Visser, 2010). It has been generally acknowledged that the organizational effectiveness and integrity of the Wehrmacht, at least on the Western Front, lasted well into the final months of 1944, even in the face of strategic defeat and staggering losses of men and materiel (Madej, 1978; Rush, 1999). When US Army psychologists Shils and Janowitz went to interview and poll German POWs captured on the Western Front between 1943–1945, they discovered that the prime factor responsible for this effectiveness and integrity was the 'steady satisfaction of certain *primary* personality demands afforded by the social organization of the army' (Shils & Janowitz, 1948, p. 281), suggesting a significant overlap between task systems and sentient systems. Although Nazi ideology and the presence of Nazi enthusiasts and party officials did play a role here (Bartov, 1991; Neitzel & Welzer, 2012), the prime factors were unit cohesion and leadership.

Regarding cohesion, what kept the ordinary German soldier fighting was 'the decisive fact that he was a member of a squad or section which maintained its structural integrity and which coincided roughly with the *social* unit which satisfied some of his major primary needs' (Shils & Janowitz, 1948, p. 284). Important here was a communality of experience, fostered by the maintenance of units to the greatest degree possible and by the replacement system, in which the 'entire personnel of a division would be withdrawn from the front simultaneously and refitted as a unit with replacements [...] [who] thereby were given the opportunity to assimilate themselves into the group; then the group as a whole was sent forward' (Shils & Janowitz, 1948, pp. 287–288; Van Creveld, 1983). This system was maintained until the very end, even to the point that regiments were allowed to become depleted in manpower by 50–75%, which depletion offset the cohesion gains (Fritz, 1996; Madej, 1978). When units were hastily formed and not properly trained at the very end of the war, group cohesion began to deteriorate (Rush, 1999).

Regarding leadership, in the German Army, to an increasing extent as the war proceeded and officer vacancies had to be filled more quickly, officers and NCOs were primarily selected on the basis of character, will power and active frontline service rather than seniority, *Stand* or General Staff experience (Knox, 2000; Van Creveld, 1983). Officers were expected to show responsibility, independent action and quick decision-making while remaining within the framework of the mission of their senior commanders. They were to lead from the front, issuing their own mission orders on the basis of first-hand knowledge of the situation. Unlike most other armies, officers were expected to live with their men and were allowed to fraternize with them when off duty. At the same time, they had to enforce strict discipline, thus combining attitudes of sternness and benevolence (Antal, 1993; Van Creveld, 1983). As one captured army officer explained:

> whether the men would follow him depended upon the personality of the officer. The leader must be a man who possesses military skill: then his men will know that he is protecting them. He must be a model to his men; he must be an all-powerful, and still benevolent, authority. He must look after his men's needs, and be able to do all the men's duties better than they themselves in training and under combat conditions. The men must also be sure that their officer is duly considerate of their lives: they must know that he does not squander his human resources, that the losses of life which occur under his command will be minimal and justified. (Shils & Janowitz, 1948, p. 297)

Probably as a result, junior officers 'were regarded by the German soldier throughout the whole Western campaign as brave, efficient and considerate', while 'senior officers, although generally esteemed, were not directly relevant in the psychological structure of the military primary group' (Shils & Janowitz, 1948, pp. 298, 299–300).

Equally important for the ordinary German soldier were the senior NCOs, 'everywhere appreciated as the most solid asset of the Wehrmacht [...] neither very interested in politics nor very aggressive, but [...] thoroughly trained, solid men' with a strong 'esprit de corps' among them (Shils & Janowitz, 1948, p. 299). The cohesion of German units was adversely affected towards the end of the war when the number of junior officers and NCOs declined and an inadequate number of replacements of lesser quality, who had fewer opportunities to connect with their men, took their place (Rush, 1999; Van Creveld, 1983).

In contrast to the German Army in World War II, the US Army had to be largely rebuilt at a quick pace when it entered the same conflict in 1941. To this end, the Americans turned to the examples of large-scale organization that they knew best, namely large corporations and the 'scientific management' that prevailed in such corporations. Traditionally assured of a strong material and technological superiority as the US was, this focus led to an army organization with a 'view of war [...] considerably more managerial than the German one, putting far heavier emphasis on doctrine, planning and control' in order to ensure the most efficient deployment of human and material resources (Van Creveld, 1983, p. 33; Schoenbaum, 1983). However, while understandable from a historical perspective, it appears that this view of war led to a much smaller overlap between task systems and sentient systems than in the German Army, as becomes clear from looking at unit cohesion and leadership in the US Army.

Regarding cohesion, the US Army employed a fixed number of 91 divisions and used replacements in men and officers to keep these divisions continuously up to strength. These replacements had to travel individually to replacement depots, then to overseas theatre depots, and from there to their divisions, a journey that took four to five months (Visser, 2010). Men and officers were then simply randomly sent to whatever vacancies existed in combat units. While administratively efficient and flexible, the steady influx of 'green' newcomers had a negative impact on unit cohesion and morale, the more so because the new men were expected to receive their advanced training from veterans. Furthermore, this whole system made the rotation of divisions in and out of the front line unnecessary and impossible, depriving veterans of the prospect of rest and recovery until they were wounded, deserted or turned into 'nervous wrecks. Perhaps more than any other factor, it was this system that was responsible for the weaknesses displayed by the US Army during World War II' (Hart, 2001; Van Creveld, 1983, p. 79).

Regarding leadership, in the US Army officers were primarily selected on the basis of intelligence rather than character, and their training was geared towards efficient management under pressure. Active front-line service did not play a role in officer selection and training, and only after 1943 was there a sufficiently large pool of commissioned officers to make it possible to rotate incompetent officers from front-line units to the rear (Visser, 2010). Officers were expected to be knowledgeable managers and loyal to their superiors, while less emphasis was put on independent action and leading from the front. Officers were not allowed to fraternize with their men, although they were expected to show them just treatment rather than enforce strict discipline. Probably as a result, '70–80 percent of all US enlisted men thought that officers put their own interests above that of their men' (Van Creveld, 1983, p. 132).

Unlike German NCOs, in the US Army NCOs were selected on the basis of seniority or ability and trained in technical skills but not in leadership skills, which were to be attained in practice. Furthermore, it was not difficult to become a NCO: in 1945 no less than 50% of all listed men were NCOs, which, together with the replacement system, created friction among the troops. With such open avenues, no 'esprit de corps' developed among NCOs (Van Creveld, 1983). It is noteworthy that after World War II, the US Army intensified its search for optimum internal efficiency in line with the principles of scientific management, which reached its culmination in the Vietnam War. During that war, the army's battlefield performance was seriously impaired by its preoccupation with administrative efficiency and quantification rather than combat effectiveness, by its officers impeccably performing administrative and procurement duties rather than leading front-line units, by its impersonal rotation system of officers, NCOs and soldiers that, as in World War II, adversely affected unit cohesion and morale, and by the resulting divide between soldiers and officers that led to mutiny and officers being killed by their own men ('fragging'). If anything, the Vietnam War represented a pointed contradiction between task systems and sentient systems that caused the total breakdown of the US Army's mental and physical integrity, as cogently pictured in movies such as *Apocalypse Now* and *Platoon* (Chwastiak, 2006; Gabriel & Savage, 1978).

The Systems Psychodynamic Perspective

The previous historical analysis can be retrospectively interpreted from a systems psychodynamic perspective. While the German Army in WWII displayed the ability to function as an integrated organization under high stress in a strategically defeated position, the US Army in Vietnam disintegrated in conditions of relatively low combat stress, with symptoms of disintegration such as desertion, fragging, mutiny and drug abuse (Gabriel & Savage, 1978). Given Menzies Lyth's hypothesis that 'the success and viability of a social institution are intimately connected with the techniques it uses to contain anxiety', the clues from the historical sources seem to indicate that the German Army was organized around the requirements of the sentient system, while these requirements were disregarded by the US Army.

This indicates that taking care of the sentient system is not just important from the perspective of the psychological wellbeing of individual soldiers. It was also a key factor in the organizations' ability to retain functional integration. The basis for the sentient system in one of the cases considered was cohesion at the operational level, which is the level most immediately confronted with the hardships of battle. Such cohesion was enabled by setting up supportive conditions such as leadership, but also by a design strategy that allowed for a certain stability in the soldiers' primary groups. While this seemed to be a main design principle behind the German Army, the US Army in Vietnam predominantly focused on task systems by designing its units along the lines of a scientific management philosophy. If a sentient system is not functioning properly, or is not even in place, organizations might develop defences against anxiety to escape from primary anxiety. According to the analysis of Gabriel and Savage (1978), it seems that this was predominantly the case for officers. They suggest that the policy of individual rotation and the fact that soldiers remained longer than officers was not an accidental by-product of an honest attempt to optimize the task system. In fact, the underlying purpose of such policies may have been to shield officers from the hazards of battle (Gabriel & Savage, 1978, p. 360). Their strongest conclusion is perhaps that the ethical and professional standards of the officer corps had decayed: leading from behind while enforcing the outward symbols of military discipline.

Quite cynically, in the US Army in Vietnam, such defences against anxiety were not available to regular soldiers. What these examples show is that while social defences may carry a negative connotation, they are not the same as a full-blown disintegration of units and organizations. If few sources to defend against anxiety are available, people might escape into strategies of psychological withdrawal. Two of the main indicators of the disintegration of the US Army in Vietnam as brought forward by Gabriel and Savage (1978)—desertion, fragging, mutiny and drug abuse—might be seen as indicative of this. Desertion quite literally amounts to fleeing from the scene and drug abuse is a prototypical example of a psychological escape. The two other indicators—fragging and mutiny—point to manifested disintegration of the army. They point to a severe hostility between groups and their leaders, and the establishment within the organization. Shils and Janowitz as well as Gabriel and Savage refer to the importance of primary group cohesion for psychological well-being and organizational effectiveness. However, not every kind of cohesion might be desirable. After all, it probably takes a cohesive group to organize a 'fragging'. Building cohesion around a shared desperation might be the ultimate perverse effect of a system that optimizes a task system from a managerial perspective.

Military Task Forces as Synthetic Organizations and Consequences for Sentient Systems

While providing insight into the various factors influencing the interplay between task systems and sentient systems, the historical examples discussed above concern standing armies engaged in full-scale combat. However, the last three decades have seen an increase in peacekeeping missions and counter-insurgency warfare. The expeditionary task forces that are formed for such missions encounter specific problems regarding the relationship between task and sentient systems. Modern military operations, sometimes referred to as military operations other than war (MOOTW; Taw & Peters, 1995), typically involve the formation of temporary expeditionary task forces that consist of many different units from standing military organizations. Some examples are recent missions in Afghanistan, Iraq and the former Yugoslavia. Military task forces involved in such missions have typically been confronted with violence, have used force themselves, have witnessed human suffering and have experienced injuries and fatalities. It is therefore also a task environment that can generate stress and anxiety, as well as, for example, moral conflicts. This indicates that a systems psychodynamic perspective might be relevant. Furthermore, temporary task forces have specific organizational characteristics that may pose significant challenges to both task and sentient systems and the interplay between the two.

According to Snook (2000), 'Task Forces are designed by taking basic unit building blocks and assembling them along hierarchical lines consistent with the demands of the mission and time-honoured military traditions of command and control' (p. 33). Other authors have related this design strategy to problems of operational flexibility (Kramer, 2007) and safety during military missions (Moorkamp, 2019). According to these studies, carried out at the Netherlands Defence Academy over a period of almost 25 years (Kramer et al., 2021; Vogelaar et al., 1996), a key characteristic of temporary task forces is the lack of a permanent organizational design. Because the military units that form the building blocks of such task forces originate from four 'parent' organizations at home (Army, Navy, Air Force and Military Police), they experience an absence of initial organizational integration and coherence. An assembly of many different units—the Dutch Task Force Uruzgan (TFU) in Afghanistan counted up to 49—has to develop an integrated and coherent task force while operating in a dynamic and dangerous mission context. A parallel can be drawn in this respect with what Thompson (1967) characterizes as the 'synthetic organization'. Based on field studies into disaster relief organizations, Thompson defines the synthetic organization as organizations that 'simultaneously have to establish structure and carry on operations' (p. 53). The choice for a 'synthetic' organization is the result of a strong demand for flexibility, which can lead to transformations in the direction of network forms of organizing.

Some studies show that innovative organizational forms emerge when task force structures are established (Moorkamp et al., 2020). Kramer et al. (2012) describe what is referred to as the smallest unit of action (SUA) in which different functional task force elements, such as infantry, intelligence and engineers, are recombined into multifunctional platoons. Moorkamp (2019) finds similar multifunctional organizational forms in combinations of flying units and ground units at the operational level. SUAs and the integration of units in the air and units on the ground showed that military personnel were trying to fundamentally shape and reshape their organization. The bottom-up design process that spawns such multifunctional organizational entities seems to provide soldiers at the operational level with more control regarding central operational processes. For example, within TFU, the combination and recombination of activities of the infantry (battle group), reconstruction teams, engineers and the cavalry within SUAs resulted in an improved ability to deal with the constantly changing demands for reconstruction in combination with the detection of improvised explosive devices and the force protection of units in the field. Similarly, combining and recombining field units with unmanned aerial vehicles and fire support in the form of Apaches, F-16s and artillery resulted in improved abilities to operate in the Uruzgan mission area. Such processes therefore seem to provide a pragmatic solution for the missing integration experienced between military units in the task force. However, our studies also show that systemic characteristics, such as the functional nature of the task force, top-down hierarchical control and the organizational complexity originating from many different units, complicated the synthesizing process. Eventually, conflicts between bottom-up design processes and obstructive systemic characteristics resulted in safety incidents between friendly units in the air and between units on the ground and those providing fire support (Moorkamp, 2019).

The Systems Psychodynamic Perspective

Our findings indicate a challenging relationship between task and sentient systems in expeditionary task forces. What emerges from the case studies is that the very flexibility that is characteristic of the synthetic organization—they require an extremely malleable task system—complicates the establishment of primary group cohesion, while this was a significant part of the sentient system in traditional armies. The case studies furthermore indicated a variation on the secondary anxieties mentioned by Menzies Lyth. During our interviews, soldiers emphasized feelings of frustration, disappointment and apathy with the military organization for its inability to provide 'solid' or 'good' ways of organizing in expeditionary mission contexts (Kramer et al., 2021). At the same time, working on pragmatic problem solving in trying to synthesize the organization was a particular source to connect task and sentient systems. As such, bottom-up self-design may create a way to deal with the inherently challenging psychological circumstances of the mission area. We see it as an important task for military management in a more general sense to facilitate such processes in the mission area.

It seems that an organization that needs to be able to continuously combine and recombine building blocks might be able to avoid social defences that are engraved in stone. This issue has not been the focus of our project, but a few interesting observations can be made. For example, Kramer (2007) argued that Dutchbat II in the 1990s—as the defenders of the Srebrenica safe area—dealt with a mission impossible by effectively ceasing to act as a crisis organization. They were essentially the hostages of surrounding Serbian forces that significantly outnumbered them. Any attempt at taking their mission seriously was met with severe intimidation, against which they were defenceless. Without reference to a systems psychodynamic perspective, Kramer observed that front-line troops appeared to turn 'inward' by, for example, obsessing about the details of planning or by taking patrol routes that would minimize the chances of encountering situations that would require intervention. A further intriguing analysis of organizing practices in MOOTW has been made by Kalkman (2019), who focused on the deployment of Border Security Teams to Chios, Greece, during the European migration crisis (2016–2017). He suggests that 'managerial actions' (p. 99) socialized front-line workers into a security frame of mind (p. 115) in which the threat to security that migrants were believed to constitute was central. He points out that the compartmentalization of activities made it increasingly complicated for front-line workers to develop an integrated understanding of local conditions for migrants (p. 115). Such managerial actions seem to shield front-line workers from difficult moral and political dilemmas. However, both examples are observations made in relation to other research questions that not systematically studied defences against anxiety. Nevertheless, they do lead to the possible hypothesis that the previously mentioned emergent process of developing innovative structures might also be influenced by psychodynamic characteristics as well as functional ones. As such, they indicate the relevance of the systems psychodynamic perspective for further research.

Conclusion

Without the benefit of specific research into the topic, this chapter cautiously explored the question as to whether the systems psychodynamic perspective might be relevant to military organizations. Given its emphasis on the importance of anxieties intrinsic to professions, one might consider its relevance to be obvious. Yet, other than the provocative reflections of historian Richard Holmes (1985, p. 236), we have found no other systems psychodynamic ideas that were applied to the military organization. Holmes suggests that military drills, and even internal bureaucracy, might not primarily have a functional purpose but, rather, a psychological one. He suggests that, given the chaos of battle, they might offer soldiers the confidence that they actually are members of a well-organized system. This might be helpful in face of anxiety, although he also suggests that they might trigger forms of escapism in the face of adverse circumstances.

The psychodynamic perspective is valuable because it directs attention to the interrelations between the psychological demands of professions and structural features of organizations. If these interrelations are ignored, organizations may be designed with a disregard for such anxieties, with potentially devastating effects, both in terms of psychological impact and the effectiveness of an organization as a task system. The exploration in this chapter leads to the conclusion that there is indeed great relevance in this perspective for the military organization. Historical examples point to the importance of establishing a healthy connection between the sentient system and the task system. Particularly in successful armies, group cohesion seems to have functioned as an important protection against the hardships of battle. Contemporary ultra-flexible 'synthetic' task forces appear to be challenged in developing such group cohesion. However, bottom-up strategies that enable the self-designing of such networks may provide the opportunity to establish a functional sentient system.

What an exploration of military cases can add to the systems psychodynamic perspective is that making the distinction between 'defence against anxiety' and 'supportive sentient system' is, at least to a degree, a moral question. Shaping military practice is an ethically relevant issue that materializes in preparing, facilitating and performing military missions (Verweij, 2009). Within that practice, people are 'moral agents' and the values they pursue are essential to the choices they make (Verweij, 2020, p. 18). The example of Border Security Teams is a case in point. At what point does protecting members of those teams against an overwhelming confrontation with human suffering become a structural arrangement that enables them to deny human suffering? Both extremes can be the cause of psychological trauma and determining where exactly this line should be drawn, whether by conscious choice or unconscious denial, is a normative question. Molendijk's (2021) study of moral injury in Dutchbat soldiers indicates the devastating psychological effects that such 'institutional betrayal' can have: if institutional realms do not face up to their own moral dilemmas, they effectively leave front-line workers to their own devices.

References

Antal, J.F. (1993). The Wehrmacht approach to maneuver warfare command and control. In R.D. Hooker (ed.), *Maneuver warfare: An anthology* (pp. 347–359). Presidio.

Bartov, O. (1991). Soldiers, Nazis and war in the Third Reich. *Journal of Modern History, 63*(1), 44–60.

Bica, C. C. (1999). A therapeutic application of philosophy. The moral casualties of war: Understanding the experience. *International Journal of Applied Philosophy, 13*(1), 81–92.

Chwastiak, M. (2006). Rationality, performance measures and representations of reality: Planning, programming and budgeting in the Vietnam War. *Critical Perspectives on Accounting, 17*(1), 29–55.

Dupuy, T. N. (1985). *Numbers, predictions & war: The use of history to evaluate and predict the outcome of battles* (rev). Hero.

Fritz, S. G. (1996). 'We are trying...to change the face of the world'—Ideology and motivation in the *Wehrmacht* on the Eastern Front: The View from Below. *Journal of Military History, 60*(4), 683–710.

Gabriel, R. A., & Savage, P. L. (1978). *Crisis in command: Mismanagement in the army*. Hill and Wang.

Gould, L. (2001). Introduction. In L. Gould, L. F. Stapley, & M. Stein (Eds.), *The systems psychodynamics of organizations: Integrating the group relations approach, psychoanalytic, and open systems perspectives* (pp. 1–16). Karnac.

Gould, L., Stapley, L. F., & Stein, M. (Eds.). (2001). *The systems psychodynamics of organizations: Integrating the group relations approach, psychoanalytic, and open systems perspectives*. Karnac.

Grossman, D. (2009). *On killing: The psychological cost of learning to kill in war and society* (Revised). Back Bay Books.

Hart, R.A. (2001). *Clash of arms: How the allies won in normandy*. Lynne Rieger.

Holmes, R. (1985). *Acts of war: The behavior of men in battle*. The Free Press.

Kalkman, J. P. (2019). *Focus on the frontline: Civil-military collaboration in domestic and European crisis management*. Amsterdam: Free University, doctoral dissertation.

Knox, M. (2000). 1 October 1942: Adolf Hitler, *Wehrmacht* Officer Policy, and Social Revolution. *The Historical Journal, 43*(3), 801–825.

Krantz, J. (2010). Social defences and twenty-first century organizations. *British Journal of Psychotherapy, 26*(2), 192–201.

Kramer, E. H. (2007). *Organizing doubt: Grounded theory, army units and dealing with dynamic complexity*. Copenhagen Business School Press.

Kramer, E. H., de Waard, E. J., & de Graaff, M. C. (2012). Task force Uruzgan and experimentation with organization design. In R. Beeres, J. Van der Meulen, J. M. M. L. Soeters, & A. L. W. Vogelaar (Eds.), *Mission Uruzgan* (pp. 235–250). Amsterdam University Press.

Kramer, E. H., Moorkamp, M., & Visser, M. (2021). Learning to organize and organizing to learn: The case of Dutch military expeditionary task forces. *The Learning Organization, 28*(3), 270–282.

Lifton, R. J. (1973). *Home from the war: Learning from Vietnam Veterans*. Other Press.
Madej, W. V. (1978). Effectiveness and Cohesion of the German Ground Forces in World War II. *Journal of Political and Military Sociology, 6*(2), 233–248.
Menzies Lyth, I. (1988). The functioning of social systems as a defence against anxiety. I. Menzies Lyth (Ed.), *Containing anxiety in institutions: Selected essays, Volume 1,* pp.43–97. London, Free Association Books.
Miller, E. J., & Rice, A. K. (1967). *Systems of organization: The control of task and sentient boundaries*. Tavistock.
Molendijk, T. (2021). *Moral injury and soldiers in conflict: Political practices and public perceptions.* Routledge.
Moorkamp, M. (2019). *Operating under high-risk conditions in temporary organizations: A sociotechnical systems perspective.* Routledge.
Moorkamp, M., Torenvlied, R., & Kramer, E. H. (2020). Organizational synthesis in transboundary crises: Three principles for managing centralization and coordination in the corona virus crisis response. *Journal of Contingencies and Crisis Management, 28*(2), 169–172.
Neitzel, S., & Welzer, H. (2012). *Soldaten: On fighting, killing and dying: The secret second World War Tapes of German POWs.* Simon & Schuster.
Petriglieri, G., & Petriglieri, J. L. (2020). The return of the oppressed: A systems psychodynamic approach to organization studies. *Academy of Management Annals, 14*(1), 411–449.
Rush, R. S. (1999). A different perspective: Cohesion, morale and operational effectiveness in the German Army, Fall 1944. *Armed Forces & Society, 25*(3), 477–508.
Schoenbaum, D. (1983). The Wehrmacht and G.I. Joe: Learning *What* from history? A Review Essay. *International Security,* 8(1), 201–207.
Shay, J. (1994). *Achilles in Vietnam: Combat trauma and the undoing of character.* Simon and Schuster.
Shephard, B. (2001). *A war of nerves. Soldiers and psychiatrists in the twentieth century.* Harvard University Press.
Shils, E. A., & Janowitz, M. (1948). Cohesion and disintegration in the Wehrmacht in World War II. *Public Opinion Quarterly, 12*(2), 280–315.
Snook, S. A. (2000). *Friendly fire: The accidental Shootdown of U.S. Black Hawks over Northern Iraq.* Princeton: Princeton University Press.
Taw, J. M., & Peters, J. E. (1995). Operations other than war: Implications for the U.S. army. *Small Wars & Insurgencies,* 6(3), 375–409.
Thompson, J. D. (1967). *Organizations in action.* McGraw-Hill.
Van Creveld, M. (1983). *Fighting power: German and U.S. army performance, 1939–1945.* Arms & Armour.
Verweij, D. (2009). Denken in dialoog: Ethiek en de militaire praktijk. *Militaire Spectator, 178*(1), 15–25.
Verweij, (Ed.). (2020). *Ethiek en de militaire praktijk.* Boom.
Visser, M. (2010). Configurations of human resource practices and battlefield performance: A comparison of two armies. *Human Resource Management Review, 20*(4), 340–349.
Vogelaar, A. L. W., Bosch, J., Kuipers, H., Metselaar, M., Nederhof, F., Witteveen, A., Van der Vlist, R. (1996). *Leiderschap in crisisomstandigheden.* FMB.

Eric-Hans Kramer is Professor in Military Management and Organization Studies at the Netherlands Defence Academy. Before that, he was associate professor Human Factors and Systems Safety at the same organization. He is involved in several research projects into organizational issues the military organization is being confronted with and with the development of research and educational programs into the psychosocial dynamics in military organizations.

Max Visser is Associate Professor of Management, Accounting and Organization at the Institute for Management Research, Nijmegen School of Management. His research is broadly concerned with relations between organizational learning, management control, accountability and performance, in particular in governmental, public sector and non-profit organizations, and from both descriptive and normative / philosophical perspectives.

Matthijs Moorkamp is Assistant Professor of Organization Design and Development at the Institute for Management Research, Nijmegen School of Management. His research focusses on organization design(ing) in risky circumstances such as military missions and crisis response. He publishes regularly on sociotechnical systems design, crisis work and military operations.

Recent Cases and Developments

Instrumental Morality Under a Gaze: Israeli Soldiers' Reasoning on Doing 'Good'

Erella Grassiani

In 2021, lethal violence against Palestinians by Israeli soldiers soared. In that year, 313 Palestinians were killed by Israel Defense Forces (IDF) fire, amongst whom many bystanders and 71 minors.[1] Israel is realizing that its soldiers are 'trigger-happy' and the IDF is doing its best to tackle this problem.[2] Interestingly, the reasons for doing so are often not based on ethical deliberations, but on instrumental ones.

For example, the Israeli newspaper Haaretz reported that 'IDF Chief of Staff Aviv Kochavi asked senior Central Command officers to take action to reduce the number of shootings of Palestinians by soldiers in the West Bank …[while] politicians and security officials criticized the conduct of Central Command chief Maj. Gen. Tomer Yadai … which they said could touch off escalation in the West Bank and hurt efforts the government is making to help the Palestinian Authority recover economically and politically'.[3] When criticizing the senior command of the IDF, these state actors did not mention that the killing of innocent people (such as non-combatants and children) was problematic in ethical terms, for its own sake. The reasoning behind reducing such lethal violence reflected the concern about the negative effects these killings could have in the shape of the violent aftermath Israel would have to deal with.

[1] https://www.btselem.org/press_releases/20220104_in_deadliest_year_since_2014_israel_killed_319_palestinians_in_opt, accessed 3 February 2022.

[2] See for example: https://www.amnesty.org/en/latest/news/2014/02/trigger-happy-israeli-army-and-police-use-reckless-force-west-bank/ and https://www.haaretz.com/opinion/editorial/to-deter-trigger-happy-soldiers-israel-must-exact-a-price-for-unjustified-killings-1.10105575, accessed 31 January 2022.

[3] https://www.haaretz.com/israel-news/.premium-idf-chief-of-staff-asks-senior-officers-to-reduce-shootings-of-palestinians-1.10102726, accessed 7 October 2021.

E. Grassiani (✉)
Department of Anthropology, University of Amsterdam, Amsterdam, The Netherlands
e-mail: E.Grassiani@uva.nl

E.-H. Kramer and T. Molendijk (eds.), *Violence in Extreme Conditions*,
https://doi.org/10.1007/978-3-031-16119-3_5

This is a clear example of what I will call here instrumental morality. Whilst used in the above by senior government officials, I have found that Israeli soldiers in the field also frequently use such reasoning when explaining their behaviour. I will argue that soldiers use such instrumental reasoning to explain, justify or legitimize their violent behaviour on three different levels, corresponding with different kinds of feelings of solidarity and responsibilities. These range from protecting the self, to the group (unit or military as a whole) to the whole nation-state. I understand instrumental morality to encompass both acts and the reasoning behind them, which are used as instruments or tools to achieve a certain goal, which is directed to the good of the self or the in-group and in the above example even the state as a whole. Such behaviour is not motivated by ethical reasoning that is guided, for example, by universal human rights or military codes of ethics. However, the end result of such acts could very well be the ethically best act, such as to stop killing innocent civilians in a situation of conflict. I explain the experiences and moral reasonings of Israeli soldiers in more detail elsewhere, where I also further explain the grounded, inductive approach that led me to my arguments (Grassiani, 2013).

The Israeli context is a specific example of 'extreme circumstances' within which this volume discusses expressions of violence by soldiers. The soldiers that feature in this chapter, whilst acting in mundane, sometimes even boring situations, are part and parcel of an extreme context that is fertile ground for extreme behaviour: the Israeli occupation of Palestinian land. The moralities of these soldiers (Grassiani, 2013) are largely influenced by this context, which is characterized by contacts with civilians, but also by the proximity between the 'battlefield' and the home front, unlike other militaries in occupying settings (US, NATO). I therefore argue that their reasoning must be analysed in the context of Israel's politics, the militarization of society and soldiers' socialization.

In contrast to mostly philosophical approaches, I argue for the use of a more social and empirical approach to morality when studying soldiers, in which social contexts and relationships that influence soldiers' reasoning and intuitive moral behaviour are taken into account (Ochs & Kremer-Sadlik, 2007). This means that in different situations and contexts, different logics, reasonings and decision-making can take place. Within the context of the military this means that soldiers' ways of thinking and their (moral) decision-making are strongly influenced by the power relations they are part of, the relationship with their comrades and commanders and the ways this context influences them physically or emotionally (Grassiani, 2013). Back home, in the context of the familiar, these same soldiers may act and reason in a very different manner. Importantly, in the Israeli context, switching from the battlefield or any other military context to one's family's home is often a weekly endeavour.

This chapter's findings are based on fieldwork I conducted for my PhD between 2004 and 2009. The approach I use for the analysis of the materials I present is based on grounded theory (Strauss & Corbin, 1997), which means the results presented here are developed 'ground up' from the empirical materials I have collected. I collected these through semi-structured interviews with a few dozen (former) combat soldiers of the IDF and through participant observations at checkpoints. All the research participants were male, and as such I will address them as 'him/his' here. I also used testimonies collected by the organization Breaking the Silence, an Israeli organization of Israeli veterans who share their stories and collect those of others about their experiences during their military service, often involving narratives of violence used against Palestinian civilians. Whilst some time has passed, I believe the narratives I collected are still relevant today, as Israeli soldiers continue to serve in the Occupied Territories and still face similar circumstances as they did then. Soldiers still man checkpoints, they still arrest people in the middle of the night and they still carry out patrols. As demonstrated in the quotes at the beginning of this chapter, the reasoning given by military personnel concerning violence by soldiers is still very similar to the reasoning I encountered then.

Some Background

All Jewish men and women in Israel are conscripted into the military and must serve from their 18th birthday; men for two and a half to three years, women for a period of 21–24 months (this has varied over the years). The soldiers who feature in this chapter and who I interviewed were all combat soldiers who served in the Second Intifada in the early 2000s. During their service, most of the work they performed resembled constabulary work, such as manning checkpoints, patrolling and carrying out arrest operations during which they interacted intensively with Palestinian civilians (Grassiani, 2013; Hammami, 2019).

Conscription is only one aspect of Israel's militarist culture, which is, for example, characterized by militarized education (Gor, 2010; Levy & Sasson-Levy, 2008, Levy et al. 2007) and which is brimming with ideas about in and out groups; who belongs to 'us' (the nation, Israel) and who does not (Ben-Eliezer, 1998; Kimmerling, 1993; Lomsky-Feder & Ben-Ari, 1999, 2007, p. 7, Kunstman and Stein 2020). The 'other' or 'them', especially for soldiers, are the Palestinians, whom they perceive as being different on almost all levels. The preservation of the self becomes critical. Instrumental moral behaviour should then be seen within the context of these in and out group dynamics.

Instrumental Morality: A Contradiction in Terms?

In order to understand the use of the concept instrumental morality in the analysis of soldiers' narratives and behaviour, we first need to grasp in more detail how a person comes to a moral act. Different theories about the role of reason and emotion within such deliberations have been developed. One example comes from Vetlesen, whose research on doing 'evil' and moral acts (1994) emphasizes the importance of emotions as opposed to seeing morality as purely rational (Vetlesen, 1994). He writes: 'moral action is logically preceded by moral judgement, that moral judgement is logically preceded by moral perception, …and the emotional faculty of empathy are equally indispensable of the exercise of moral perception and moral judgement' (1994, p. 350). Vetlesen also writes that 'emotions anchor us to the *particular* moral circumstance … to the *here and now*' (Ibid., p. 4, italics in original). He thus explains that before we make a moral judgement and perform moral actions, our emotions lead us to assess the situation we find ourselves in (moral perception).

Whilst it is important to take the role of emotions into account when speaking about morality and when looking at moral behaviour and moral judgement, it is equally important to consider the processes of reasoning and subsequent acts that accompany these emotions. Important for the case at hand, such processes of *thinking about* or reflecting on one's actions can sometimes take place during the actions, but mostly happen after these actions have been carried out. In theories on ethics, reasoning means using ratio and being conscious and detached about one's decisions and actions. This would exclude intuitive, unconscious moral behaviour and activities (Haidt, 2001). Thus, as both Vetlesen and Haidt propagate, we also need to take into account intuition and feelings that influence behaviour in a split second without elaborate reasoning taking place at that very time. I will demonstrate the importance of considering the circumstances our research participants are in at the time of the events and acts they describe to us and the feelings they mention. Further, the reasoning that takes place should also be contextualized.

Reasoning, from this perspective, does not necessarily mean that a very long process of detached thought precedes every activity described. By the time a research participant gave me his answer during my fieldwork, there is a chance he had in fact thought and reasoned about it. However, not all actions taken by soldiers are the result of *extended* reasoning; the physical and especially emotional state the soldiers are in and the relationship they do or do not have with the persons in front of them strongly influences their ad-hoc decision-making as well (Grassiani, 2013). Most of my research participants shared with me their reasoning as it was at the time of the events they were describing, without whitewashing their behaviour after the fact. As such, I accept their explanations, even post-hoc, as the way they interpreted their behaviour at the time. The terms 'reasoning' and 'action' or 'behaviour' are used here and are part of morality, depending on the context within which an action took place and the way the actor speaks about it.

What Is Instrumental Morality?

The 'instrumental' in 'instrumental morality' means that acts and reasoning about those acts is geared towards an end goal that is beneficial to the self (or the group). What can make such an act 'moral' in the eyes of the self (or the group) is the fact that the result of the act falls within the socially or institutionally agreed upon 'doing good' within the specific context the act falls in. In order to analytically categorize an act or reasoning as instrumental morality it is of importance to establish the motivation for the act, its end goal and the context of the act. In their work on business policy, for example, Quinn and Jones look at the moral obligations of managers (1995) and differentiate between different motivations such managers have for their actions. In the case of managers who act in an instrumentally ethical way, they 'might do what is morally proper, but they do so to increase shareholder wealth' (Quinn & Jones, 1995, p. 23).

In my previous work (Grassiani, 2013), I argued that a similar point can be made for Israeli soldiers serving in the Occupied Palestinian Territories. Their actions vis-à-vis the Palestinians they encounter can appear to be moral or 'right' in the eyes of the (international) public or according to military ethics; however, often the motivations that triggered them are not 'principled' but serve to achieve a goal or agenda that is directed to the good of the self or the in-group. Here the primary consideration is not the ethics of the act itself, but what could be achieved by it in favour of the actor or the in-group. This, however, does not mean that motivations are always egocentric or unjust.

Importantly, the 'other' who is necessarily present when we speak of moral acts, is not or hardly considered here, although he or she is affected, sometimes even positively, by such instrumental moral behaviour. From the outside, such behaviour seems to be on a par with moral values; people are given help and seemingly treated with respect. The reasoning behind such behaviour, however, can make it instrumental if it is self-centred and general universal rights are not taken into consideration.

Scrutinizing instrumental morality gives us the chance to understand more about actual motivations for actions and lets us look beyond the surface of mere observation. Furthermore, behaving morally out of instrumental considerations means that in instances where no one is watching, immoral behaviour is more easily displayed. The importance of a 'gaze' when talking about instrumental ethics is considerable (Grassiani & Verweij, 2014). Without a watchful (external) eye, one's performance will be different (depending on who is present, and thus who is possibly an 'internal' eye) and motivations for acting in specific ways change.

What Is in It for Me: The Personal Level

When an actor acts on the basis of the consideration of what is in it for him or her, actions become instrumental on a personal level. To accomplish maximum personal gain through action, the concept of obedience is important for soldiers. In this case, the fear that is linked to disobedience is a motivating force, for if one does not obey his superiors one could be punished. Soldiers will try to avoid this punishment and 'save themselves' at all times. In other words, obedience becomes instrumental to them.

This could also be linked to the issue of discipline within the military. Soldiers are taught to behave according to strict rules and their behaviour is 'on display' and under scrutiny almost at all times. This 'disciplinary gaze' (Grassiani & Verweij, 2014) of the military, then, is an important and powerful factor that influences soldiers' behaviour. Below I will also discuss another goal soldiers seek to achieve on the personal level, which is the desire to stay human, or, as soldiers put it, 'to keep their human dignity'.

Avoiding Punishment

Many soldiers stated that avoiding being reprimanded by a superior, or worse, receiving actual punishment for one's deed, is a motive for action. This is closely related to the issue of obedience within the military and the power it has over soldiers. The fear of being caught disobeying a superior can be a strong motivator. This becomes clear from the answer Barak gives when asked about the code of ethics and the way he as a soldier was informed about it:

> Only in basic training, it's bullshit, you don't use it. Like purity of arms, you know that if you forget your gun your officer will leave you on Saturday in the base so you take your weapon. You know that if you act to the Arab with too much violence the officer will tell you, they don't say do whatever you want. Me and my friends, we never hit a person at a check post or something, just like that. Never happened. One time when we found that bomb, the guy started to be wild and we had to keep him quiet.[4]

What immediately stands out—besides Barak's dismissal of the code of ethics as 'bullshit'—is the motivation he gives for behaving in a correct way: he wants to avoid criticism or punishment by his superiors. The fear of being reprimanded by one's superiors is one of the dominant motivators for soldiers' actions; behaving according to the books or strictly following the orders of commanders is then often instrumentally motivated.

Furthermore, such behaviour is performed to impress the commander and thus needs to be *seen* and is usually displayed when the commander is present. This makes the behaviour instrumental, I pose, as it is not performed out of a belief that it is the only right or just thing to do or because a code of ethics prescribes it, but to impress others or to avoid their disapproval. In a way, soldiers present an 'ideal', humane soldier to their commanders. Adam, a company commander himself, emphasizes the extent of the commander's influence:

> I think that what interests the soldiers is not doing it [hitting] in front of the company commander, more than not to hurt Arabs. He will prefer that the Arab will sit another hour in the heat or cold and wait for the company commander then that he would decide himself. At the end of the day he stands in front of the commander, he takes care of the promotions, if he goes home. It's his mother and father. Like in school, don't tell my father.

In this case not hurting a Palestinian, which seems like morally just behaviour, is motivated by fear of the scrutiny of the commander, not by the realization that it is (morally) wrong. We could say that this soldier has internalized the disciplinary gaze of the military mentioned earlier. What Adam also conveys here is the fact that soldiers, at times, find it hard to decide and prefer for a Palestinian to suffer for a

---

[4] All interviews were carried out in Hebrew and translated by the author.

bit longer than to make a wrong decision for which they could be punished. The company commander has a great deal of power over the soldiers, he is their 'mother and father' and hence his influence on the behaviour of soldiers and their decision-making is great. Such understanding should ideally result in proper moral behaviour in the eyes of these commanders. However, the reasons given by commanders for behaving in a particular way, not only to avoid punishment, are often instrumental by nature as well, as we shall see shortly. The understanding instilled in soldiers does not guarantee a less instrumental approach.

Keeping My Human Dignity

A different motivation to act morally is related to an idea many soldiers mentioned, namely to keep 'our own humanity' and 'our own human dignity'. Within such discourse, the soldiers were not so much concerned with the wellbeing of the other, in this case the Palestinian, but instead with their own 'saneness', the saneness of the soldiers themselves, the preservation of their intrinsic human properties i.e. being a moral human being. Eviatar, in a fragment of the interview that was quoted earlier, puts this very clearly:

> …If we need to check a car, we will get everything out of the car. The question is how you do it. You don't throw anything, you don't start messing in his stuff but you ask the person to take the stuff out of the car. Maybe it does not interest the person if you do it in a polite way, but it's more to keep our *tselem enosh* (human dignity).

This preoccupation with the moral wellbeing of the self, the effort to keep the moral characteristics of oneself intact, I argue, is a case of instrumental morality. Soldiers who reason in this fashion have a clear goal in mind, a goal that is related to their own feelings of worth and not to the wellbeing of others. In this case, Eviatar wants to act 'correctly', one could say ethically, not because the person in front of him is a human being and deserves to be treated well, but because he wants to continue to feel he himself (Eviatar) is still a human being. Golan told me a striking story of an arrest operation after which he found himself in the back of the military vehicle with an arrestee. He noticed they were both born in the same year and tried to make conversation through an interpreter:

> When I was sitting with the terrorist, it was very important for me to have some kind of contact, maybe from naivety, I was determined to stay a human being and not become a machine of hate and fear.

As becomes clear from his words, this effort to make conversation with the Palestinian man sitting next to him handcuffed and blindfolded has a clear goal, namely for Golan to feel like a human being and not become 'a machine'. This conversation was not initiated to understand the arrestee, for example, but for the soldier to feel he was doing a good thing as a human being.

Protecting Your Brothers

In the case of an instrumental discourse on the level of the group or institution soldiers are part of, their reasons for making decisions or behaving in a certain way (or for refraining from certain behaviour) take the best interest of the group into consideration. On this level, the motivation behind the behaviour of soldiers could for example be to avoid 'giving the military a bad name' or to hide certain activities from public scrutiny.

Protecting Your Soldiers or Comrades

> The soldiers must want his Regiment, his comrades and those around him to survive. The Regiment is his family, where he is not alone…the Regiment provides the opportunity for him to become the best soldier in the world; he fights for something more than himself; he fights for his comrades and the regiment; and indirectly, for his home and his family. (From a Canadian military document in Winslow, 1997: 74)

Feelings of solidarity and comradeship are deeply interconnected with military life and the motivations of soldiers to go into combat. Grossman quotes Dyer who writes that this strong feeling of accountability towards your comrades is 'a special kind of love that has nothing to do with sex or idealism' (1995, p. 150). Whilst some argue that this comradeship does border on sexual desires (Kaplan, 2002; Sasson-Levy, 2008), Dyer rightly states that feelings between soldiers could be called a special kind of love, which goes very deep and is immensely important for soldiers.

Many works on the military have emphasized such strong feelings and most authors have found that the motivation to fight first of all comes from the will not to let down your buddies (Grossman, 1995; Holmes, 1985). Having a group around you makes the meta-motivations as ideological ideas less important, because what you do, you do for the comrades you share your direct experiences with.

An important motivation for the action or inaction of soldiers is the fear that they or their comrades will get hurt. When assessing an operation, it is often not judged by its moral standards, but by how dangerous it is for the force, thus for the soldiers themselves. The ways in which the behaviour of soldiers or commanders can be geared towards their own group, their soldiers' or their comrades' safety, can be made out from Dror's words. Here is a commander who clearly draws the line when it comes to the safety of his soldiers:

> In the reserves what I always say; don't know if it's more or less important, do me a favour: take care of yourself. That's the most important, you have to do the job but … and here I'm ready to pay the price of warmth and humanity, I'd rather have my soldier in one piece than a Palestinian that's crying. This cold decision like I told you before.

When any kind of group is very close-knit, its behaviour and reactions towards the outside world can become very harsh in an effort to protect their in-group, whilst disregarding the safety of the 'other'. As Lomsky-Feder and Ben-Ari write 'the Arab is the enemy against which one is fighting, and the "other" through which Jewish Israeli identity in general and that of the warrior in particular are defined' (2007, p. 7). Importantly, situations as described above are most often not life or death situations, but daily interactions with Palestinians that are, from the point of view of the soldiers, fraught with uncertainty.

Protecting the Image of the IDF

'He [the soldier] won't shoot in the mass, if besides this Arab with a M16 he shoots another 10 Arabs, we would look very bad. The soldier wouldn't want that as his responsibility' (citation from interview).

Another issue I came across in my interviews with Israeli soldiers was a strong awareness of the image of the IDF. It seems that they are very conscious of how the world sees the IDF and, by extension, sees them. As already touched upon before, soldiers would behave 'properly' when in sight of others not connected to the IDF. Not only for their own good, but also for the good of the IDF. They made an effort to protect the IDF from external criticism, avoiding the staining of its name. Snir gives a good example of how easily the military and its soldiers can get a bad name through misunderstandings on the side of the outsiders, undeserved in his opinion:

> You can't always tell what you are doing. There is a big fear in the IDF today to talk to the media. You are not allowed to talk to the media, in any case. They can also hurt their company. For example the papers write about a soldier of Givati in Gaza, just the soldier started laughing or something, they take it out of proportion, so they protect their own company. For example a soldier can say to his Ethiopian friend that he has been sleeping in the same room with for 3 years, "*Ya kushi*"[5] as a joke. A journalist hears that, immediately sensation. So it's not only fear of the military, but also it's about practical things. .... "they call Ethiopians "*kushim*" the battalion commander comes and takes the company apart, sends this guy to jail.

The point Snir is making here is the fact that the outsider, the media in this case, can hurt the soldiers because they misinterpret their 'joking' and alert the public, which can lead to a reaction by the higher commanders, such as isolating the soldiers from their comrades or even arresting them. This is the reason soldiers, in his opinion, should refrain from speaking to the media, as this could 'hurt their company'. I argue that these soldiers have internalized the power of the disciplinary gaze of the media. Nir, a commander from the *Nahal* battalion explains that he and his comrades would be careful when searching Palestinian houses and give their motivation for being careful: they learned '...not to break anything, so they won't accuse us later that we destroyed things. We would never touch, that's it'. Nir emphasizes that he and his comrades would be careful not to break anything so 'they', which could be their superiors or external observers such as human rights organizations, would not criticize or punish them for it.

[5] *Kushi* is a derogatory name for blacks in Israel, usually Ethiopians, comparable to the N-word in English.

Another motivation for soldiers' actions related to the protection of the good name of the military, is to avoid looking 'weak' in the eyes of the Palestinians, in line with military strategy. Looking weak could mean risking attacks or losing the upper hand in the power dynamics that are present between Palestinian civilians and Israeli soldiers. Dror, who was a commander with the paratroops and who was very motivated whilst serving in the IDF, talks about a dilemma he faced, which will further clarify this point:

> One of my dilemmas was, and that's why I didn't like to do checkpoints, because it's a case…you really meet the population, not the terrorist. Now go and be selective, on the colour of eyes, if he has a beard, he's aggressive, he's not, he is thin, he is fat. The dilemma that I had was, you think about the slap that the commander gave or the soldiers gave, not that it happened to me but I heard it from stories. You wonder if this slap came to the person who deserved it…or you slapped someone who believed in the checkpoint or in the IDF, he didn't like it but somehow he believed and no…think about the little boy that saw his father being slapped…. it's done, you can't work on that anymore. And there are those commanders, they keep children close to the soldiers, they know they don't like the field food (*khamgashiot*) and they would pass it to the children. Now go figure if you give it to a child that thinks these soldiers are "sissies" (*freierim*) or maybe you gave it to a child that will grow up and would respect, most or a lot of the Palestinian population has respect for the IDF. I don't know if this is because of the media or because they met the IDF and saw a person that treated them with respect.

The dilemma described here can be summed up as follows; once you hit someone or give a kid some extra food, for that matter, you do not know what the effect of this action will be, what the consequences of your actions are. You could appear to be weak, for example, when treating the population with a soft hand. On the other hand, if you treat the population harshly and aggressively, you might generate more hatred. The intention behind both behavioural patterns, 'soft' or aggressive is, however, the same: to gain a strong, good image of the IDF in the eyes of the Palestinian population and to reduce feelings of hatred and thus possibilities of attacks.

Guy found himself in a difficult situation as a commander with the paratroops, in which he had to make a swift decision that could have led to an innocent person's death, but could also have saved the lives of others. He observed a Palestinian man with a donkey and a large gas canister. His initial thought was that it could be the attacker they were on the lookout for and that this was a bomb the man was carrying. When he looked more closely, he saw an old man walking along with a young boy, probably his grandson, taking a gas canister to his home:

> So this thing about making the decision, it's crazy….we saved our good name, he got away…and lucky that I got so much balance to decide here is this old man with the gas balloon that could be a bomb, if I would have shot him, his grandchild would…I, from my point of view, see it as a circle; the grandfather falls next to the grandchild, he'll be a terrorist, terrorist, he'll be my enemy. That was the dilemma that was post trauma as I call it now. That was one of the things that influenced me most.

The way he reasoned about the situation is interesting and very telling; the fact that he made the right decision and did not kill the old man 'saved our good name' protected the good name of the unit. Furthermore, if he had killed the innocent man, his grandson, in Guy's opinion, would have become a terrorist. Thus, he does not only think about the good name of his unit, but also about the creation of hate and possible terrorists.

Do Not Get the State Involved

'The orders are first of all to take care of yourself. To take care not to hurt, not to make a mistake… and get the state involved' (quote from interview).

Concerning the final level that I want to discuss here, the reasoning and acts of the soldiers and commanders are characterized at a more macro level, for the good of the state; the security of the state and its image in the eyes of outsiders, particularly the international community. An example of such reasoning is, for example, a soldier who is careful about shooting into a crowd in the event of riots, because the death of innocent people is bound to have political consequences. More important than not killing an innocent person is the political implications such an action could have for the state. Yossi, a commander who was active in combat during Operation Defensive Shield gives a good example:

> This was actually the complication of the combat in all the refugee camps. I think that apart from Bethlehem and Jericho and Gaza, I was in all the cities and this was actually the motto of the combat; to know that you are fighting among civilians, and to think a thousand times before you shoot for the first time. Because the first shot can lead to a catastrophe. A catastrophe, if you take down (*morid*) a child or if you take down a mother, it can be such a chaos that it is not in your league.

Whilst I do not wish to say here that this commander does not care about killing a child or a mother, his first concern is the chaos it would create, a situation that is 'out of the league' of a soldier. Another example of such reasoning comes from Adam, also a commander:

> It has to go to the lower level. In my time, the IDF used to say that what one soldier does, can influence a whole country. Because if one soldier, for example, just shoots for no reason, or would make a mistake it can go to the state. If someone innocent gets killed, we know where it will be tomorrow. It does not stay with the soldier. That's why it's important that everything gets down to the level of the soldier.

In this quote, Adam explains why, in his eyes, it is important to forward operational information down to the 'normal' soldiers in the field and not to leave them in the dark about what is happening around them and about why this is happening. The reason he gives, and which as he says was the message of the IDF as a whole during the time he served, is that the consequences of the actions of a soldier can influence something bigger, a whole country. In other words, if a soldier makes a wrong move, this could, for example, severely impair political negotiations or ties with neighbouring countries. In recent years, the IDF has seen several of such cases in which the media got a hold of footage of soldiers mistreating or even killing people in violation of its code of conduct.

The military's incentive for easing the life of Palestinians is ultimately, according to this example, the greater security of the region and thus the state of Israel. Hence, I argue it has an instrumental character. Offer explains the decision-making processes during Operation Defensive Shield and the way in which a bad image of Israel was created in the eyes of the world:

> So afterwards…there were discussions. To blow up, not to blow up (*lefotets*) [houses]. On the other side if you blow up (*mefotets*) the whole world shout, "what, you blew it up". Also in Jenin, even though I wasn't there but our unit took part in the refugee camp there, and I believe every word they tell me…afterwards, an Arab director made a film about it, that we did there, I don't know what.[6] In Europe they said we…it was the case that they [soldiers] surrounded from all direction and they closed in. The terrorists didn't flee, they couldn't flee, there is a ring. They closed in all the time, until there were 10 houses, in a refugee camp, everything is closely built, what are you going to do with that ... go in with soldiers and have a face-to-face fight? So that you will have 30 killed and they will have 30 killed, that's not …with all respect to them, our lives are more important than theirs. Not more important, but from our point of view more important. From the point of view of the Heaven it's the same human being. If I'm in a war, I prefer him to get killed then that I would get killed or my friend. That's the way it is, he also prefers that an Israeli get killed then a Palestinian. So, they [IDF] went in with bulldozers, D9, and they destroyed all the houses. What can you do, they told them [Palestinian militia] to surrender, 2 days, 3, 4 days, [and they stayed inside]. So the state of Israel did a massacre and killed poor children and there were snipers, and 6 years old kids came out and they hit their heads. There is no such thing; it's not in the education of the military. But that's what the world sees, for some reason.

Offer's analysis is very telling. It is clear to him that through the media and, for example, through documentaries, the international community gets the wrong picture about the activities of the IDF and therefore Israel gets a bad name. By explaining how the IDF worked and showing the lack of choice the militants gave the military (it's them or us), he tries to show how this negative image, which is a fabrication according to him, was created. Such a negative image of the IDF, but also of the state as a whole, is something soldiers are taught to avoid, hence a ban on speaking to journalists and other outsiders who could potentially cause harm through their 'twisting of the story', as we also saw before. Great care is thus taken by soldiers to avoid misunderstandings about their activities and their behaviour is adjusted accordingly.

Conclusion

In this chapter, I have outlined the instrumental morality that resonates in Israeli soldiers' discourses. I defined instrumental morality as seemingly moral behaviour towards the 'other', motivated by goals directed to the good of the self or the in-group.

[6] The film Offer mentions here is the documentary 'Jenin, Jenin' by Muhammad Bakri, a Palestinian-Israeli filmmaker. This film caused a lot of commotion in Israel because of its critical stance towards the activities of the IDF. The film was banned from Israeli cinemas and Bakri was even taken to court by a group of reservists who served during ODS.

I divided the moral behaviour and reasoning I encountered in my research into three different levels, as the motivation of soldiers was not only geared towards an ego-centric goal, but was often also triggered by feelings of solidarity with comrades or the defence of the state.

On the personal level, these soldiers tried to avoid punishment from superiors, to stay human beings, to avoid becoming 'machines' and most of all to try to keep safe. On this level the reasoning and actions of soldiers within the OPT were directed towards their own good, their own wellbeing.

On a group level, soldiers felt they needed to protect their comrades, just as commanders have strong feelings of responsibility for the wellbeing of their soldiers. Action was then taken for the benefit of the safety of the in-group. Furthermore, efforts were made to protect the image of the IDF. Good behaviour is displayed when outsiders such as human rights organizations and reporters are present to avoid giving the IDF a bad name.

On the more abstract level of the state, both soldiers and commanders made the same efforts. The realization is that if a soldier makes a mistake, this can become a political scandal and influence the precarious political position of Israel within the Middle East. 'Good' or 'just' behaviour by soldiers and commanders is therefore often displayed to avoid scrutiny and repercussions on different levels. The other, the Palestinian citizen, however, is hardly considered.

Crucial to the instrumental morality I have described is that it exists when it is observed by someone. The gaze of the superior in the military setting proved to be very influential for the way soldiers behaved, as they wanted to behave according to what they thought their superiors expected. Another disciplinary gaze that proved to be important was that of the international community or the press, which disciplined the soldiers into behaving as they thought they should (morally speaking) in order to avoid the above-mentioned consequences.

References

Ben-Eliezer, U. (1998). *The Making of Israeli Militarism.* Indiana University Press.

Gor, H. (2010). Education for war in Israel: Preparing children to accept war as a natural factor of life. In K. Saltman & D. A. Gabbard (Eds.), *Education as enforcement: The militarization and corporatization of schools.* Routledge.

Grassiani, E. (2013). *Soldiering under occupation: Processes of numbing among Israeli soldiers in the Al-Aqsa Intifada.* Berghahn books.

Grassiani, E., & Verweij, D. (2014). The disciplinary gaze of the camera's eye: Soldiers' conscience and moral responsibility. *Journal of Military and Strategic Studies, 15*(3), online.

Grossman, D. (1995). *On killing: The psychological cost of learning to kill in war and society.* Back Bay Books/Little.

Haidt, J. (2001). The emotional dog and its rational tail: A social intuitionist approach to moral judgment. *Psychological Review, 108*(4), 814–834.

Hammami, R. (2019). Destabilizing mastery and the machine: Palestinian agency and gendered embodiment at Israeli military checkpoints. *Current Anthropology, 60*(S19), S87–S97.

Holmes, R. (1985). *Acts of war: The behaviour of men in battle.* Cassell.

Kaplan, D. (2002). *Brothers and others in arms: The making of love and war in Israeli combat units*. Harrington Town Press.
Kimmerling, B. (1993). Patterns of militarism in Israel. *European Journal of Sociology, 34*(2) 196–223. https://doi.org/10.1017/S0003975600006640
Kuntsman, A., & Stein, R. L. (2020). *Digital militarism.* Stanford University Press.
Levy, G., & Sasson-Levy, O. (2008). Militarized socialization, military service, and class reproduction: The experiences of Israeli soldiers. *Sociological Perspectives, 51*(2), 349–374.
Levy, Y., Lomsky-Feder, E., & Harel, N. (2007). From "obligatory militarism" to "contractual militarism"—Competing models of citizenship. *Israel Studies, 12*(1), 127–148.
Lomsky-Feder, E., & Ben-Ari, E. (Eds.). (1999). *Military and militarism in Israeli Society.* SUNY Press.
Lomsky-Feder, E., & Ben-Ari, E. (2007). *Identity, politics and the military in contemporary Israel* (Unpublished manuscript), Hebrew University, Jerusalem.
Ochs, E., & Kremer-Sadlik, T. (2007). Introduction: Morality as family practice. *Discourse & Society, 18*(1), 5–10.
Quinn, D., & Jones, T. (1995). An agent morality view of business policy. *Academy of Management Review, 20*(1), 22–42.
Sasson-Levy, O. (2008). Individual bodies, collective state interests: The case of Israeli combat soldiers. *Men and Masculinities, 10*(3), 296–321.
Strauss, A., & Corbin, J. M. (1997). *Grounded theory in practice.* Sage.
Vetlesen, A. (1994). *Perception, empathy, and judgement: An inquiry into the preconditions of moral performance.* Penn University Press.
Winslow, D. (1997). *The Canadian airborne in Somalia: A socio-cultural inquiry*. Canadian Government Publishing.

Erella Grassiani is Associate Professor at the Department of Anthropology at the University of Amsterdam. Her work traces the flows of the (Israeli) security worldwide and looks at the way cultural ideas, technologies and consultants move around globally. In the past she has done extensive research on the Israeli military. She is the author of *Soldiering under Occupation processes of Numbing among Israeli soldiers in the Al-Aqsa Intifada* (2013 Berghahn Books).

Soldiers as Street-Level Bureaucrats? Military Discretionary Autonomy and Moral Professionalism in a Police Perspective

Teun Eikenaar

> We are doing some good work for these people, but I joined the Army to be in a combat-ready unit, not to be a policeman
>
> American peacekeeper. (Kitfield, 2000, in: Franke, 2003, p. 34)

From Military Stereotypes to A Revaluation of Soldiers' Discretionary Autonomy

Traditionally, if there is one organization that is associated with obedience, rule compliance and an alleged lack of discretion on the part of its personnel, it is the military organization. From stereotypical depictions in popular culture to a plethora of academic analyses, all point to the military as a command and control organization that revolves around obedience to authority (Kucera & Gulpers, 2018, p. 366). Furthermore, military occupational culture has often been described in terms of a standard model with recurring characteristics of hierarchy, rules, discipline and obedience (Soeters, 2018).

These ideas about traditional military organizational and cultural structures are said to have a considerable relevance for military ethics. For instance, Olsthoorn et al. (2013) state that the military is distinguished from other organizations by its specific organizational circumstances, in which especially the loyalty to direct colleagues 'is the part of the military ethic that is most at odds with what a "regular" professional ethic entails' (2013, p. 86). Kucera and Gulpers (2018) frame this military ethic in terms of two dominant orientations on moral responsibilities: military ethic seems to revolve largely around the inward orientation on fellow soldiers, and the upward orientation on the political community or state, whereas the outward orientation on

T. Eikenaar (✉)
Radboud University, Nijmegen, The Netherlands
e-mail: teun.eikenaar@ru.nl

E.-H. Kramer and T. Molendijk (eds.), *Violence in Extreme Conditions*,
https://doi.org/10.1007/978-3-031-16119-3_6

'significant others' (allies, enemies, civilians) is mainly conceived of in terms of negative moral obligations, i.e. soldiers are expected to minimize harm to civilians.

In this contribution, I suggest that this traditional imagery has been superseded and, most importantly, that a 'constabularization' of the military (Easton & Moelker, 2010; Janowitz, 1964) necessitates a revaluation of military discretionary autonomy and moral demands. In other words, I explore the possible moral implications of the partial shift of military work towards police-like work. Quite a few studies point to the fact that battlefields, missions and military organizations have changed, but only a few elaborate on what these changes might mean for how soldiers have to navigate new discretionary and moral demands. I reflect on the possible implications of a change in discretionary autonomy by first mapping some important changes in military organizations and missions and their possible moral implications. Next, I introduce some of the moral demands police officers face as 'street-level bureaucrats' and discuss how these demands might increasingly apply to military settings. By comparing changes in military work with police officers' discretion and the moral character of that profession, I hope to provide a particular perspective on what it might mean to be a contemporary soldier.

Military Autonomy: From Effective Fighting Power to New Moral Demands

Military autonomy understood as room for the decisions and initiatives taken by soldiers and lower ranking officers is in no way a new concept. However, relatively recent (post-Cold War) developments have placed the discretion of soldiers in a new perspective and in a more complex occupational reality. I will first outline how military autonomy might be read as the result of a change in command philosophy. Next, I will indicate what recent changes might mean for this autonomy and (consequently) for the moral demands placed on military personnel.

The notion of autonomy has quite a long tradition in military thinking. Especially in German military theory, going at least as far back as the writings of Clausewitz, and through the army reforms of General Von Moltke, the ideal can be observed of a 'professional officer corps with the authority and willingness to take decisions in real time at a low level' (Bungay, 2005; Van Bezooijen & Kramer, 2015). This ideal was grounded in military strategical considerations: the 'fog of war' leads to a degree of uncertainty that demands 'individual initiative, independent thinking and responsibility at all levels of command (Visser, 2017). As such, operational autonomy—known in this context as *Auftragstaktik* as opposed to *Befehlstaktik* (Bungay, 2005)—was given credit for an instrumental reason, namely to build 'a more effective fighting organisation' (Visser, 2017). These views on fighting power became a military tradition and later contributed to the successes of the German army during World War II (Bungay, 2005; Wilson, 1989).

The German notion of *Auftragstaktik* directly inspired ideas of 'mission command'. As an alternative to centralized forms of command that see initiative as a form of disobedience, the notion of 'mission command' advocates empowerment of those at the forefront of events (Bungay, 2005; Van Bezooijen & Kramer, 2015). In this view, military personnel on all levels are expected to make decisions, guided by more generic descriptions of mission goals (cf. De Graaff & Kramer, 2012, p. 44). It paved the way for an instrumental view of military autonomy; that is, freedom was granted to use one's own discretion—in line with the intentions of those higher in command—so that military objectives would be easier to achieve (Bungay, 2005).

Military autonomy understood in these terms is grounded in a specific conceptualization of war and how it can be fought effectively. However, the context in which ideas of *Auftragstaktik* came into being has since changed substantially. These changes have both increased the dependency on military discretionary autonomy and added other moral dimensions to it.

After the Cold War ended, a more complex geopolitical interplay of forces has led to, among other things, a partial change from 'threat-driven' to 'ambition and capability-driven' military forces, a general downsizing of the military and the deployment of military forces for crises, 'postmodern conflicts' and new missions (Hoffenaar, 2017; Kaldor, 2007; Manigart, 2018). In this respect, traditional wars have given way to 'new wars', in which states disintegrate and the distinction between combatants and non-combatants disappears, as does that between legitimate violence and criminal behaviour (Kaldor, 2007). These wars demand a different approach, one in which individual rights are given prominence over state rights; in other words, a human security approach (ibid.). Elsewhere, changes have been described mostly in terms of a shift from 'combat-related deployment' towards 'peacekeeping' (Franke, 2003). Following Franke (2003), a growing array of peace operations (military operations other than war, or 'MOOTW') is generating debate on how to differentiate between 'humanitarian assistance, preventive diplomacy, peace-making, peace-keeping, peace-building, and peace enforcement' (ibid., p. 33). Such differentiation is beyond the scope of this contribution, but it is clear that military work has gone through a significant evolution.

These changes have several important consequences for the organization of the military. One of the most conspicuous is the changing outlook of military missions. Instead of traditional warfare with large, opposing land forces, present-day missions are intended to be relatively quick, flexible and temporary deployments in which multinational cooperation takes place through task forces (De Graaff & Kramer, 2012). Some claim that these general changes have led to a further 'contraction of command', as a result of which the traditional military hierarchy has flattened (Manigart, 2018). As such, this further underlines the philosophy of 'mission command' mentioned above, albeit in part for a different reason. Mission command here is not only the result of thoughts on effectiveness, but also of circumstances in which centralized command and control is not always easy to maintain. The 'dynamic complexity' of contemporary missions implies a degree of ambiguity, indistinctiveness and equivocality that makes it impossible to centrally plan and control military operations in their entirety (Kramer, 2007). Thus, military autonomy is not (only) the result of an instrumental military tactic to enhance fighting power, but also the unavoidable outcome of how missions have changed.

This also leads to an increase in responsibility for mission leaders and front-line personnel, and requires a great deal of improvisation and ad hoc cooperation (Kramer & Moorkamp, 2019; Moorkamp et al., 2016; Richardson et al., 2004). Soldiers are expected and obliged to learn and adapt *during* their missions, to deal with uncertainty and to cooperate and achieve mission goals while improvising. As such, military organizations appear to be losing some of their characteristics as 'total institutions' in which all behaviour and roles are predictable (Goffman, 1961).

Most importantly for the point made here, these changes have a relevance for the moral demands on soldiers. Military organizations have long actively refuted the adage that 'orders are orders', and soldiers are expected to independently judge how to act in specific situations on the basis of their own moral considerations (Verweij, 2002). However, it seems that these moral obligations are being made more complicated by the aforementioned changes and associated responsibilities (Richardson et al., 2004). Under the circumstances described above, soldiers are perhaps more than before confronted with the moral question of having to judge when and how to apply force (Van Baarda & Verweij, 2006; Verweij et al., 2007), especially when they have to balance this with the expectations of involvement, respect and insight into local communities (see also Kucera & Gulpers, 2018). Moreover, if 'human security' is key in these 'new wars' (i.e. the security of individuals and communities instead of state security; cf. Kaldor, 2007), soldiers are confronted with the question of how this security is locally constituted, what actors influence it and how (with what means) it might be enhanced. All of these aspects imply (moral) demands that soldiers will have to deal with. As such, this human security agenda has important consequences for what is expected from military personnel. Among other things, communication, consultation and dialogue are key aspects of this role in new wars: a bottom-up approach that goes beyond 'winning hearts and minds' towards truly gaining insight into local communities and their needs (Kaldor, 2007).

To an important extent, these changes and the ensuing demands on military personnel point to matters of discretion. Military personnel are expected to decide in conditions that are often largely unknown to them, and are handed a significant amount of discretion that they might not be expecting or trained for. Moreover, they have to do so in contexts in which peacekeeping is considered more important than combat.

Discretionary Autonomy and Moral Demands: The Case of the Police

A way to describe these changes and to understand their implications for the military profession is by comparing contemporary military work with police work. I argue that police officers' discretionary autonomy is of a specific *moral* nature and that an exploration of this area might help shed light on possible moral demands in contemporary military frontline work.

As early as 1964, Janowitz described the military establishment as a 'constabulary force when it is continuously prepared to act, committed to the minimum use of force, and seeks viable international relations, rather than victory' (Janowitz, 1964, p. 418). Others literally frame recent military developments as a 'politicization' of military work (Kucera & Gulpers, 2018). During these deployments, it is not the absolute use of force but practical conflict resolution that counts (Easton & Moelker, 2010; Janowitz, 1964), and conduct that resembles community policing is expected of soldiers (Easton & Moelker, 2010). This is also evident in the composition of human security forces, which consist of mixed units of military, civilian and police forces that in fact resemble police units more than traditional military units (Kaldor, 2007).

As a result, frontline military personnel might increasingly be seen as 'street-level bureaucrats' (Lipsky, 2010) or 'streetcorner politicians' (Muir, 1977). They have to deal with complex situations fraught with impactful decisions in which it often does not come down to the use of force as such but instead to the question of how to limit the use of coercion to a minimum and solve issues by other means. The aforementioned 'human security agenda' also clearly brings to mind the principles of specific forms of policing, in particular community policing, as this too is centred around the priorities and (feelings of) insecurity of citizens (Skogan, 2009; Terpstra, 2010). These settings call for other competencies and capabilities and another type of moral professionalism than has hitherto been expected from soldiers.

The question, then, is what are these moral demands that police officers face and that might be instructive for the military profession? Although it is widely accepted that all street-level work can be framed in terms of moral reasoning and that frontline professionals use their own 'normative order' for judging the people they encounter in their jobs (Maynard-Moody & Musheno, 2003, p. 158), police work is known to be of a *particular* moral nature; that is, through their relationship with the use of force, police officers operate in a unique moral space with unique moral demands in which their discretionary autonomy in decisions concerning the use of coercion is a vital and inherent part of their professionalism. In this respect, a police officer might see restraints upon his discretion as hampering his 'capacity to fulfil his assigned task' (Skolnick, 1966, p. 90). Some therefore state that the question of what authority he formally has is less important than the constitution of his authority through his uniform and the way he uses this in the street. A police officer sees criminal procedures through the eyes of a 'craftsman' and emphasizes his own expertise to judge on appropriate measures (Skolnick, 1966, p. 196). In other words, it is inherent to police officers' work to steer on the basis of what Frans Denkers et al. (2001) call their 'moral compass'. Therefore, it is an *illusion* to fully discipline police officers' behaviour, because, 'police officers to a large extent act on their own terms. In fact, they have to, due to the nature of their work. […] police officers remain reliant on their moral compass' (Denkers et al., 2001, p. 37—translation my own). There is a rich intellectual and academic tradition that shows that their professional behaviour as 'moral entrepreneurs' in practice is more important than an application of legal and formal rules (Denkers et al., 2001; Herbert, 1996; Manning & Van Maanen, 1978). As such, Denkers et al. (2001) point out that morality is at the heart of police work by stating, among other things, that the police are defined by the *expectation* that they are able to deal with injustice in a decent and non-biased way, and that they *factually* deal with the vast majority of offences through various moral practices rather than by invoking criminal law or through law enforcement (Banton, 1964; Bittner, 1967; Denkers et al., 2001; Skolnick, 1966; Terpstra, 2010).

A work that addresses these moral practices in detail is William Muir's, 1977 study *Police: Streetcorner Politicians.* In this book, Muir specifically addresses variations in how police officers work by posing the central question 'What makes a police officer good?' and mapping various police work styles. Muir does so by defining two virtues of police work. The first one centres around the use of legitimate force. Muir explains this as the extent to which police officers know how to resolve the contradiction between coercive means and just ends. In Muir's view, this can be done in either an 'integrated' way in which the use of coercion is part of 'an overall moral code', or in a 'conflictual' way in which the application of coercion is likely to cause guilt 'because it is not related to basic moral principles' (ibid.). Other authors have equally recognized this matter as inherent to police discretion. In fact, this particular moral challenge sets police officers apart from other frontline professionals: having the possibility to legitimately use force if needed comes with specific (moral) responsibilities and expectations, and even defines police officers *as* police. Police officers are the ultimate authority to be called upon when force is needed, *and* they are expected to use it in a decent way (Bittner, 1970; Muir, 1977; Terpstra, 2010; Van den Brink et al., 2016). A second reason why police officers' use of discretion comes with particular moral questions follows from what Muir calls the challenge of how to grasp 'the nature of human suffering'. Although Muir understands this as an intellectual endeavour, it can also be understood in moral terms. In his view, police officers approach human suffering either 'cynically' or 'tragically'. Officers with a cynical approach frame people in binary terms such as good or bad/us and them and police work is about fault-finding. When police officers look tragically upon human suffering, they support a more unitary vision of mankind and 'action [is] complexly produced by chance, will and circumstance' (Reiner, 2010, p. 133).

Muir understands the responses to these two virtues to boil down to a fourfold typology. He therefore identifies the moral heart of the profession by mapping how police officers respond to the moral challenge of how to apply coercive means and how to grasp (and, might I say, morally judge) human suffering. This typology can be found in similar terms in many other studies on police work styles (Reiner, 2010; Van den Brink et al., 2016). Here I use Reiner's overview of Muir's categories:

> The "avoider" (with cynical perspective and conflicted morality) shirks duties; the "reciprocator" (tragic perspective and conflicted morality) hesitates to use coercive power even when appropriate; the "enforcer" (cynical perspective and integrated morality) acts in the heat of conflicts and without understanding the need for restraint; the "professional" (tragic perspective and integrated morality) is the "good" cop. He or she is able to use violence where necessary in a principled way, but is adept at verbal and other skills that enable solutions to be resolved without coercive force wherever the opportunity exists. (Reiner, 2010)

In conclusion, morally challenging interactions or moral dilemmas appear to be inherent to police officers' work because of the very nature of that work.

These conceptualizations can also be interpreted in normative terms. To develop a notion of what is a 'good policeman', Muir uses Weber's model of a 'mature man' (1977, p. 54) and his view of a 'professional politician' with both 'passion' and 'perspective' (and here it becomes apparent how Muir leans on Weber's normative idea of a professional politician). In Weber's view, passion is the 'capacity to integrate coercion into morals'. In other (and Muir's) words, a truly professional police officer has 'passion', which implies that he has an integrated morality and sees the use of coercion as part of an encompassing moral code and as part of his profession. Conversely, the lack of passion points to a 'conflicted morality' and thus objections to coercive means, as these do not fit with an officer's basic moral principles. In addition, Weber's notion of perspective refers to an intellectual objectivity; to 'the knowledge of tragedy with which all action, but especially political action, is truly interwoven' (Muir, 1977, p. 51). Thus, this Weberian perspective comes down to the 'tragic perspective' described above. Conversely, a lack of such perspective comes down to a cynical worldview that justifies violence out of a lack of conscience and ethical concerns of civilization. In sum, Muir uses the Weberian notion of a professional politician to define a 'good policeman' as someone who 'felt morally reconciled to using coercion [i.e. having passion/an integrated morality—TE] and at the same time […] reflected empathetically upon the condition of mankind [i.e. having perspective/a tragic perspective—TE]' (ibid., p. 54). In terms of the typology used above, a good policeman (what Muir calls 'the professional') is someone who has both an integrated morality and a tragic perspective.

Soldiers as Street-Level Bureaucrats?

The question is, what is the relevance of this analysis for (new) military contexts? In this section, I try to answer that question in three parts: (1) does a constabularization of military missions mean that soldiers develop similar (moral) work styles?, (2) what could this analysis mean for military moral professionalism?, and (3) to what extent are military (organizational) conditions suitable for embedding possible lessons from such a comparison?

One question that follows from the above exploration is the following: can the police work styles described above be used to understand responses of military personnel to moral demands? After all, although Muir appears to have specific (normative) ideas about the 'true police professional' (see below), his model is primarily grounded in an empirical analysis: all four work styles can be found among police officers.

For one thing, several authors have noted that various social identities permeate military praxis and point out that these identities or roles are uncorrelated or even conflict (see for instance op den Buijs et al., 2019; Franke, 2003; Janowitz, 1964). Franke (2003), for instance, points to the opposition between warrior and peacekeeper roles and presents an analysis of four cognitive strategies soldiers might employ to navigate and resolve the resulting cognitive inconsistencies between these roles (ibid., p. 41). For the point made here, Franke's fourth model is most interesting: the 'post-Cold War military professional' in which 'the peacekeeper subidentity [is viewed] as an integral part of [soldiers'] professional self-conceptions rather than merely as "a job that it takes a soldier to do"' (ibid., p. 46).

However, as Franke himself also notes, the integration of a 'peacekeeper subidentity' into a soldier's self-image does not resolve the moral dilemmas that come with it, and does not clarify moral demands and issues of moral judgement. As such, analyses such as these do not go much further than the oppositions between different 'cognitive frames' or 'roles', and, in Franke's case, stop at describing a mere two roles. Interesting as these may be, these roles seem rather reductive of the present-day military praxis. As such, they say but little about moral dispositions and interpretations of discretion. Moreover, cognitive strategies say little about how soldiers deal with changing (moral) circumstances. In this respect, the exploration presented here might also serve as a suggestion for a new framework to (empirically) explore contemporary military discretion and work styles.

Another question that follows from these explorations is a normative one: can this typology be used to reflect on and maybe improve moral professionalism among military personnel? In this connection, Muir labels only one of the four work styles as 'professional'. The others are thus 'nonprofessional policemen' (Muir, 1977, p. 55). Interesting in this respect is that other authors apply the label 'professional' to other work styles. Reiner (1978, in: Reiner, 2010), for instance, uses the same two virtues to devise a fourfold typology, but reserves the label professional for those police officers with a tragic perspective but a *conflicted* morality (what Muir calls a 'reciprocator'). Hence, officers who hesitate to use coercion, even when appropriate, are considered by Reiner to be 'professionals'. All in all, both these 'virtues' invite reflections on soldiers' moral professionalism.

First of all, if soldiers carry out more police-like work, it might mean a different relation to the role of coercion. Although soldiers cannot be seen as killing machines that blindly follow orders, and moral considerations on the application of violence are part and parcel of the military profession, new types of missions and police-like work imply a specific relationship to violence. As such, 'traditional', stereotypical notions of military work might be in line with an 'integrated morality' towards coercion in which coercion is an integral part of an overall moral code. Perhaps new police-like deployments also require work styles with a more conflicted morality and military personnel with a different moral attitude towards the use of violence, especially since the use of force is less central. As such, the refuted claim that the uniqueness of the military professional resides in his expertise in 'war-making and in the organized use of violence' (Janowitz, 1964, p. 15) has still not led to the ultimate conclusion, namely that the military profession might be less and less about the use of force and that this might call for a different military moral professionalism. Consequently, dealing with problems of 'human security' and gaining insight into local communities and their needs might require military personnel whose professionalism is also defined by a conflicted morality towards violence. This calls for a new perspective on military moral professionalism.

Secondly, new types of missions might also have consequences in terms of the second virtue described above; that is, the extent to which frontline personnel grasp the nature of human suffering. If new missions increasingly revolve around 'significant others' who cannot and should not be reduced to enemies, a thorough rethinking of what Kulcera and Gulpers (2018) dub the outward orientation of the military ethic may be needed. Where they make a case for 'humanitarizing' the military, another way of framing this imperative is by subscribing to what has above been named the 'tragic perspective'. This does not stop at 'more respect' for others outside the organization (ibid., p. 369). It might go as far as meaning that the intellectual grasping (and moral judging) of the world outside the gates of the compound should be inherently tragic. Thus, an outward ethic that is truly humanitarian perhaps starts with a tragic foundation and the acknowledgement of 'a sense of the meaning of human conduct – a comprehension of the suffering of each inhabitant of the earth, a sensitivity to man's yearning for dignity, and, ultimately, "some kind of faith" that no individual is worthless' (Muir, 1977, p. 51). For the military, the challenge might be obvious here, especially since the 'cynical perspective' and its common 'intellectual grasping' of enemies and civilians as savages is all too familiar to many soldiers (Bandura, 2002; Ivie, 1980).

A third matter for consideration is whether present military conditions are suitable for organizationally embedding such views on military moral professionalism. For one thing, within the police, there are many forces that hamper moral professionalism and restrict police officers in the daily execution of their jobs. As early as 1966, Skolnick already noted how police reforms tended to bureaucratization and reduced police work to a machine-like execution of predictable tasks. Such bureaucratizing tendencies have received ample academic attention since then. Also in recent years, police organizations in multiple countries have apparently tended to curtail police discretionary autonomy (Terpstra et al., 2019).

In view of the foregoing, an interesting question is how developments towards politicization and the associated possible changes in discretionary autonomy, moral demands and the moral professionalism of soldiers fit within a military organization that still fosters traditional ethics and expects obedience and loyalty. It would appear that soldiers have to face new demands while still operating within an institution that seems in part to be stuck in traditional structures. For instance, new expectations are at odds with the control exercised by higher ranking officers (Manigart, 2018, p. 421), and soldiers are still by and large trained along the lines of their functional units, whereas ad hoc mission compositions require that they think outside of that particular box (De Graaff & Kramer, 2012; Kramer & Moorkamp, 2019). It appears that these structures can lead to ambivalence in practice, and soldiers are expected to be able to cope with confusion and ambiguity (Manigart, 2018, p. 420). Some authors frame this lack of clarity and preparation and training for situations that require improvisation in terms of a 'double bind': servicemen and women are expected to follow orders but are at the same time required to improvise and to be creative (cf. De Graaff & Kramer, 2012; Kramer & Moorkamp, 2019).

This double bind may only become worse when soldiers experience a need for more discretion, while they work for an organization that in spite of recent changes maintains elements of a 'greedy institution' with high demands on commitment (Coser, 1974). In this respect, it remains to be seen how a traditional morality of seclusion and loyalty will fit with new types of work that require discretionary autonomy, openness, interaction, a different relationship to violence and a different perspective on civilians. In addition, the notions described here can help to create a more versatile image of military frontline work, but this obviously has important implications for training, recruitment and command structure, as 'there will need to be an additional "broadening" of the education base of the military establishment in order to equip military professionals with expertise in areas beyond core war-fighting skills and to develop a capacity for flexibility in these turbulent New Times' (Dandeker, 1994; cf. Kucera & Gulpers, 2018). As such, a rethinking of military morality also requires a rethinking of how this morality can be instilled through training, supported and developed.

Conclusion

The purpose of this contribution was to provide a specific perspective on developments in the military profession. Basically, this contribution consisted of a reflection on what the consequences of a politicization of military work might be for military discretionary autonomy, moral demands and moral professionalism.

Resolving the tensions this chapter opened with perhaps starts with truly acknowledging the discretion needed in contemporary military missions. Part of this acknowledgement concerns the new police-like moral demands that come with them *and* the provision of the training and organizational conditions needed for the development towards a new moral military professionalism. In other words, if being a soldier increasingly resembles being a police officer, it should also be acknowledged that this implies having to negotiate situations that require other forms of moral thinking, judging and acting.

Obviously, it remains to be seen what the exact implications of these developments will be for (moral) professionalism within the military. A detached reflection like the one presented here raises more questions than it provides answers. An adaptation of the military ethic might be necessary to gear soldiers to human security tasks, but what does a further politicization mean exactly for selection, education and training? What does it mean for recruitment? How do you present working for the military, and what sentiments do you trigger when you want to draw the right military professional, if this 'good soldier' should resemble a 'good police officer' with a 'tragic perspective' and an 'integrated morality', or maybe actually a 'conflicted morality' (cf. Franke, 2003; Janowitz, 1964)? What does it mean for military organizations when various moral demands and expectations have to be balanced (the 'traditional inward and upward ethics' in case of combat-ready warrior roles versus the moral demands of police-like forms of military work)? Can these various ethics be part of one organization at all? And of one ethic?

A key conclusion *does* perhaps follow from what Franke (2003, p. 45) calls the transcending or integrative model to resolve the tension between warrior and peacekeeping roles; that is, instead of trying to deny the peacekeeper role, this role could be accepted as being a part of the 'post-Cold War military professional'. First and foremost, this means that military socialization and training should contribute to the acceptance of multiple roles as integral parts of the professional identity. Furthermore, such training could prepare soldiers for the moral challenges, moral judgements and moral entrepreneurship that are all a part of military discretionary autonomy—without having the pretence that anyone other than the member of the military on the front line can meet these moral demands.

References

Bandura, A. (2002). Selective moral disengagement in the exercise of moral agency. *Journal of Moral Education, 31*, 101–119.

Banton, M. (1964). *The policeman in the community*. Tavistock.

Bittner, E. (1967). The police on skid-row: A study of peace keeping. *American Sociological Review, 32*, 699–715.

Bittner, E. (1970). *The functions of the police in modern society: A review of background factors, current practices, and possible role models*. National Institute of Mental Health, Center for Studies of Crime and Delinquency.

Bungay, S. (2005). The road to mission command. *British Army Review, 137*, 22–28.

Coser, L. A. (1974). *Greedy institutions: Patterns of undivided commitment*. Free Press.

Dandeker, C. (1994). New times for the military: Some sociological remarks on the changing role and structure of the armed forces of the advanced societies. *The British Journal of Sociology, 45*, 637–654.

De Graaff, M., & Kramer, E. H. (2012). Leiderschap, uitzending en 'intelligent failure'. De intelligente mislukking als hoeksteen van de lerende organisatie. *M & O, 5*, 41–60.

Denkers, F., van Beers, P., & Elzenga, W. (2001). *Frans Denkers' Moreel kompas van de politie*. Ministerie van Binnenlandse Zaken en Koninkrijksrelaties, Directie Politie; Koninklijke De Swart [distr.].

Easton, M., & Moelker, R. (2010). Police and military: Two worlds apart? In M. Easton, M. Den Boer, J. Janssens, R. Moelker, & T. Vander Beken (Eds.), *Blurring military and police roles*. Eleven Publishing.

Franke, V. C. (2003). The social identity of peacekeeping. In T. W. Britt & A. B. Adler (Eds.), *The psychology of the peacekeeper: Lessons from the field*. Praeger.

Goffman, E. (1961). *Asylums: Essays on the social situation of mental patients and other inmates*. Aldine.

Herbert, S. (1996). Morality in law enforcement: Chasing "bad guys" with the Los Angeles Police Department. *Law & Society Review, 30*, 799–818.

Hoffenaar, J. (2017). De Nederlandse krijgsmacht in historisch perspectief. In E. R. Muller, J. M. J. Bosch, I. M. De Jong & P. A. L. Ducheine (Eds.), *Krijgsmacht. Studies over de organisatie en het optreden*. Wolters Kluwer.

Ivie, R. L. (1980). Images of savagery in American justifications for war. *Communication Monographs, 47*, 279–294.

Janowitz, M. (1964). *The professional soldier: A social and political portrait*. Free Press.

Kaldor, M. (2007). *Human security: Reflections on globalization and intervention.* Polity Press.
Kramer, E.-H. (2007). *Organizing doubt: Grounded theory, army units and dealing with dynamic complexity.* Liber & Copenhagen Business School Press; Marston Book Services.
Kramer, E.-H., & Moorkamp, M. (2019). Reflective practice in synthetic expeditionary task forces. In W. Klinkert, M. T. I. B. Bollen, M. Jansen, H. De Jong, E.-H. Kramer, & L. Vos (Eds.), *NL ARMS Netherlands Annual Review of Military Studies 2019: Educating officers: The thinking soldier—The NLDA and the Bologna Declaration.* Asser Press.
Kucera, T., & Gulpers, L. (2018). A military ethic for new missions. In G. Caforio (Ed.), *Handbook of the sociology of the military.* Springer International Publishing.
Lipsky, M. (2010). *Street-level bureaucracy: Dilemmas of the individual in public services.* Russell Sage.
Manigart, P. (2018). Restructured armed forces. In G. Caforio (Ed.), *Handbook of the sociology of the military.* Springer International Publishing.
Manning, P. K., & Van Maanen, J. (1978). *Policing: A view from the street.* Goodyear Pub. Co.
Maynard-Moody, S., & Musheno, M. C. (2003). *Cops, teachers, counselors: Stories from the front lines of public service.* University of Michigan Press.
Moorkamp, M., Wybo, J.-L., & Kramer, E.-H. (2016). Pioneering with UAVs at the battlefield: The influence of organizational design on self-organization and the emergence of safety. *Safety Science, 88*, 251–260.
Muir, W. K. (1977). *Police: Streetcorner politicians.*University of Chicago Press.
Olsthoorn, P. H. J., Bollen, M., & Beeres, R. (2013). Dual loyalties in military medical care—Between ethics and effectiveness. In H. Amersfoort, R. Moelker, J. Soeters, & D. Verweij (Eds.), *Moral responsibility & military effectiveness* (pp. 79–96). T.M.C. Asser Press.
op den Buijs, T., Broesder, W., Goldenberg, I., Resteigne, D., & Kivirähk, J. (2019). Warrior and peacekeeper role identities: Associations with self-esteem, organizational commitment and organizational citizenship behavior. *Journal of Military Studies, 8*, 3–15.
Reiner, R. (2010). *The politics of the police.* Oxford University Press.
Richardson, R., Verweij, D., & Winslow, D. (2004). Moral fitness for peace operations. *Journal of Political & Military Sociology, 32*, 99–113.
Skogan, W. G. (2009). *Police and community in Chicago: A tale of three cities.* Oxford University Press.
Skolnick, J. H. (1966). *Justice without trial: Law enforcement in democratic society.* Wiley.
Soeters, J. L. (2018). Military culture. In G. Caforio (Ed.), *Handbook of the sociology of the military.* Springer International Publishing.
Terpstra, J. (2010). *Het veiligheidscomplex: ontwikkelingen, strategieën en verantwoordelijkheden in de veiligheidszorg.* Boom Juridische uitgevers.
Terpstra, J., Fyfe, N. R., & Salet, R. (2019). The abstract police: A conceptual exploration of unintended changes of police organisations. *The Police Journal: Theory, Practice and Principles, 92*, 339–359.
Van Baarda, T. A., & Verweij, D. E. M. (2006). *Military ethics: The Dutch approach—A practical guide.* Nijhoff.
Van Bezooijen, B. J. A., & Kramer, E. H. (2015). Mission command in the information age: A normal accidents perspective on networked military operations. *Journal of Strategic Studies, 38*, 445–466.
Van den Brink, G., Hendrikx, W., Van Hulst, M., Maalsté, N., & Mali, B. (2016). *Een onderzoek naar de morele weerbaarheid van Nederlandse politiefunctionarissen.* Boom Criminologie.
Verweij, D. (2002). The dark side of obedience: The consequences of Hannah Arendt's analysis of the Eichmann case. *Professional Ethics, 10*, 143–158.
Verweij, D., Hofhuis, K., & Soeters, J. (2007). Moral judgement within the armed forces. *Journal of Military Ethics, 6*, 19–40.
Visser, M. (2017). Teaching giants to learn: Lessons from army learning in World War II. *Learning Organization, 24*, 159–168.
Wilson, J. Q. (1989). *Bureaucracy: What government agencies do and why they do it.* Basic Books.

Teun Eikenaar (PhD) is a (postdoctoral) researcher with a background in anthropology and philosophy. His interests and research concern (uniformed) frontline workers, moral professionalism, and issues of public safety and (plural) policing. Currently, his research focuses on the subject of moral injury among police and military workers in relation to organizational circumstances.

The Future of the Comprehensive Approach as a Strategy for Intervention

Marenne Jansen and Eric-Hans Kramer

The tragic events in Kabul in August 2021 marked the end of the US mission and 'western' military intervention in Afghanistan, but arguably also marked the end of the 'comprehensive approach' as a functional model for foreign intervention that integrates security and development into a peacebuilding strategy. In the conceptual realm, doubts regarding this approach long predated the scenes of August 2021. During the Bonn conference in 2001—at a moment in which Afghanistan was still in internal conflict—it was agreed that the country had to be economically and politically rebuilt while a UN mandated mission (UNAMA) would assist the Afghan authorities by keeping Kabul and its surroundings safe. In hindsight, it seems to have been an overly optimistic and maybe even naive plan to rebuild and develop a country that is still in conflict. To start understanding these events and their implications for international politics, it is relevant to take a closer look at the thinking behind international interventions over the past twenty years, and the alleged relationship between security and development.

Grossly simplified, the relationship between security and development is defined by the question to what extent development can be achieved without a secure environment, and to what extent a level of development is needed to establish a secure environment. More concisely, it is about 'the struggle for priority between poverty alleviation and safety concerns' (Glasius, 2008, p. 40). Although perspectives on this relationship have differed over time and per context, the general mantra for most missions at the beginning of this century has been: 'There will be no development without security and no security without development' (Annan, 2005). Missions that

M. Jansen (✉)
Head Strategic Unit, Ministry of Defense Netherlands, Hague, The Netherlands
e-mail: mm.jansen.02@mindef.nl

E.-H. Kramer
Faculty of Military Sciences, Netherlands Defence Academy, Breda, The Netherlands
e-mail: fj.kramer@mindef.nl; erichans@xs4all.nl

E.-H. Kramer and T. Molendijk (eds.), *Violence in Extreme Conditions*,
https://doi.org/10.1007/978-3-031-16119-3_7

adhere to this mantra are built upon a combination of security forces and development. They bear names such as the 'comprehensive approach', 'whole of government approach' or '3D (Defence, Diplomacy and Development) approach'.

This chapter evaluates the application of the comprehensive approach in the last decades by placing it against the background of the human security concept. Human security was introduced in the 1994 UNDP report on human development with the intention to broaden the concept of security and not only focus on the military protection of territories of sovereign states (AIV, 2019). As such, human security is a concept that underpins and justifies the comprehensive approach. In this chapter, three different interpretations of human security are identified. These interpretations serve as an analytical framework to discuss if, and if so how, the comprehensive approach has a future as a strategy for intervention.

To set the stage for analysis, the comprehensive approach and the experience with its implementation is discussed below. What follows is first an elaboration of the comprehensive approach followed by a discussion of the underlying security-development nexus. Subsequently, three interpretations of human security are introduced, which set the stage for the evaluative perspective on the implementation of the comprehensive approach. In conclusion, it will be argued that a viable future for the comprehensive approach depends on which interpretation of human security is dominant at an institutional level.

The Comprehensive Approach

The term 'comprehensive approach' was developed against the background of the major crises of the 1990s: the end of the Cold War, the failed UN peacekeeping operations in Somalia, Rwanda, and Bosnia-Herzegovina and subsequent doctrinal change in the nature of peacekeeping missions (Chandler, 2012; Gabrielse, 2007; Gammer, 2013; Rietjens & Bollen, 2008; Travers & Owen, 2007). It is an approach that searches for the integration of a variety of rebuilding initiatives that reflect the combination of security and development as a peacebuilding strategy. While a universal definition is hard to find, there is a shared belief that governments, donors, international organizations and civil society all play a role in seeking to address and prevent conflict, and often struggle to do so in a complementary and coordinated way. Therefore, this approach aims to combine developmental work, political support and security strategies, which makes the case for combining civil and military efforts. As such, the comprehensive approach is aimed at facilitating system-wide coherence across the security, governance, development and political dimensions of international peace and stability operations (Chandler, 2001; De Coning & Friis, 2011; Van der Lijn, 2011).

The comprehensive approach seeks 'to achieve greater harmonization and synchronization among the activities of the international and local actors, as well as across the analysis, planning, implementation and evaluation of the program cycle'

(De Coning & Friis, 2011, p. 3). The terminology used by governments and multilateral organizations, such as the UN and the NATO, differs. A variety of interpretations of the comprehensive approach exists, reflecting governmental and institutional preferences at national levels and in different international institutions. The approach has, for example, been used in policymaking (Briscoe & Van Ginkel, 2013; Clinton, 2009; Commission on Human Security, 2003) practiced by the UN, EU, AU, the World Bank, other international organizations, national governments, local and international NGOs and has been widely studied (Chandler, 2012; Frerks & Klein Goldewijk, 2007; Glasius, 2012; McCormack, 2011; Tschirgi, 2005). National implementation and policy-making on the comprehensive approach dates back to around 2003–2006 (mainly Sweden, the UK and Denmark). The Table 1 shows which organization uses what term, and indicates the particular (sub)system of local and international actors that these terms are used by. As shown, there are nuances and slightly different accents. However, in general it is safe to state that terms such as 'the whole of government approach', the '3D approach' and the 'integrated approach' can all be grouped under the umbrella of the comprehensive approach.

Internationally, the concept was established during the Riga summit in 2006 and applied from 2006 onwards (ISAF). The UN established the concept as the guiding principle for complex future post-conflict operations, stating that 'Integration is the guiding principle for the design and implementation of complex UN operations in post-conflict situations and for linking the different dimensions of peacebuilding (political, development, humanitarian, human rights, rule of law, social and security aspects) into a coherent support strategy'. Furthermore, the EU and NATO adopted the comprehensive approach concept to describe their respective initiatives to pursue coherence (IRSEM, 2010). The EU developed its approach in the wake of the Cold War, the Balkan crises, the 9/11 attacks and the subsequent operations in Afghanistan.

The wide usage indicates that the comprehensive approach is attractive, while at the same time the variety in emphasis indicates that it is complex. It is a broad term that most actors will initially feel at ease using. Complexity starts when actual cooperation, coordination, integration or co-existence of development, diplomacy and development needs to be established on the ground. Several authors have formulated critique on the way the comprehensive approach has been implemented (Glasius, 2012; Paris & Sisk, 2009; Schirch, 2011; Van der Lijn, 2011). Such critique emphasizes, for example, the degree to which local ownership is truly achieved on the ground, or the degree to which coherence between different actors in the host nation and the national levels of contributing nations is maintained. These are issues of implementation. However, the conceptual basis of the comprehensive approach itself is also worth discussing, which we do below.

The Security-Development Nexus

In the introduction, it was argued that the comprehensive approach can be better understood against the background of what is called the security-development nexus.

Table 1 Modified from De Coning and Friis (2011, pp. 3–6)

Actor	Preferred term	Meaning/Context	Focus
UN	Integrated approach	Type of operational process and design Mainly about the different elements of the UN family integrated into a single country	*Intra-agency* A guiding principle for future post-conflict operations and complex peacebuilding missions. It functions on the Security—Development nexus
NATO	Comprehensive approach	Since NATO, as a military organization, is incapable of achieving a system-wide effect, NATO can solely be regarded as a contributor to the CA of the wider international community	*Internal–External* The relationship between NATO, as a military organization, and the other international actors engaged in the same theater
EU	Comprehensive approach	To pursue coherent policies, bring together the different tools and capabilities of EU policy, such as European assistance programs, the European Development Fund and the Member States' military and civilian capabilities	*Inter-agency* Civil-military relationship among the elements of its crisis management approach, i.e. the military, the rule of law, protection and conflict management
National level	Comprehensive approach Whole of Government approach 3D approach	Institutional approach (UK, US, Canada): Dedicated units, typically housed in the MfAs to manage the whole of government systems	*Whole of government* Coherence among government departments and agencies of a specific country
		Process approach (Netherlands, Norway, Sweden): Based on a white paper-level requirement for policy integration, backed up by series of meetings at the ministerial and technical levels	

To clarify this, it is necessary to discuss why development and security are connected in the first place. Although today the relationship between security and development is a common topic on the agenda of international security, in the past the relationship between these two concepts was rarely considered, as the concepts originate from different domains. Security was believed to be the concern of states—referring to enemies, territorial control and armed coercion. In contrast, the domain of development was dominated by phrases such as foreign aid, poverty, education, health care, infrastructure and governance (Hettne, 2010). While during the Cold War security analyses had been primarily focused on interstate conflict between two superpowers, the period after the Cold War required a perspective on new multilateral peace operations in which more complex and less familiar tasks had to be performed (Johnstone et al., 2005, p. 57; Paris, 2004, p. 17). Particularly UN peacekeeping missions in the 1990s (in for example Somalia, Rwanda, Bosnia-Herzegovina, East Timor, Kosovo) indicated a need for a new type of peace operation. After the emergence of intrastate conflicts and associated threats in the 1990s, the focus in political, military, humanitarian and academic realms shifted (Frerks & Klein Goldewijk, 2007, p. 23). Later, the Brahimi report (UNSC, 2000) also emphasized the need for operations in which interrelated political, economic, and developmental as well as security problems were to be addressed simultaneously, with a focus on people, as well as states.

This shift from a narrow military focus to a more integral view meant that security became just one aspect of a more extensive process. Consequently, the traditional peacekeeping paradigm changed drastically. When peacekeeping operations changed in nature, the dominant doctrine for international peace and stability became peacebuilding. Peacebuilding is more civilian than military in content, as it emphasizes the combination of political and development activities targeted at the sources or causes of conflict. In peacebuilding missions, the involvement of multiple actors (UN, NATO, regional organizations, development agencies, (international) non-governmental organizations) is required to perform more complex and less familiar tasks (Paris, 2004, p. 17). Peacebuilding interventions require a combined, coherent approach, including strategies from the realms of both security and development. More specifically, peacebuilding requires approaches that combine interventions in a coherent, i.e. comprehensive, way. This rapprochement of civil and military actors in crisis management forms the basis of the comprehensive approach.

This explanation makes clear that the development of the comprehensive approach required the adoption of a security-development nexus, which in the 1990s was triggered by the historical events in the international arena. Already at that time, these ideas were connected to an underlying concept of 'human security'. Human security is generally recognized as a 'people-centered, comprehensive, context-specific and prevention-oriented framework through which national capacities could be strengthened' (UN Secretary General, 2012, p. 4). Essential in this view is that no longer states, but individuals are the prime actors, perpetrators and victims in this new type of violence (Kaldor, 2007). It became current in international policies and challenges the traditional agendas of power, particularly because it places the needs of the individual—not the state—at the center of the security perspective.

The concept was introduced in the UNDP report of 1994. The UNDP report notes that: *'human security can be said to have two main aspects. It means, first, safety from such chronic threats as hunger, disease and repression. And second, it means protection from sudden and hurtful disruptions in the patterns of daily life-whether in homes, in jobs or in communities. Such threats can exist at all levels of national income and development'* (UNDP, 1994, p. 47). The later, broad definition is often summarized as 'freedom from want'; the first, as 'freedom from fear'. This difference represents a difference in understanding of what human security could be. It appears that a subtle emphasis is put on either the security or the development part of the security-development nexus. Notwithstanding such nuance, the shift towards human security represents a significant shift in point of orientation and therefore also a shift in commitment in international politics. This indicates that the development of the comprehensive approach required (1) establishing the conceptual relation between security and development, i.e. the security-development nexus, and (2) the shift in international relations from an emphasis on state security to human security.

While the postulate of a security-development nexus has not been very controversial, the concept of human security has. One of the issues is that the concept is used for different purposes. Sometimes human security is used strategically, on other occasions it is used in a more activist framework, and yet others use the concept in what can be called an idealistic interpretation. Some see it as a new paradigm, others as a continuation of an old state-centered vision of security. Some use it as a potential basis for action on the ground; others apply it solely as a policy discourse. Some will claim that this vagueness is intentional, in order to enable all to agree on something, others see the potential of and continue to express the need for clarification. Such critiques focus on potential intentions behind human security and therefore also on the very integrity of—comprehensively approached—missions that are conducted in its name. These differences in perspective on 'human security' are significant, because they indicate that the justification for a comprehensive approach does not inevitably follow from an unequivocal definition of 'human security'. Therefore, depending on a particular perspective, efforts at organizing 'comprehensively approached missions' could be evaluated quite differently. The next section offers a framework to categorize different positions in relation to human security.

Perspectives on Human Security

In this section, we describe three appreciations of human security. These three appreciations refer to different ways of understanding political intentions behind the concept and, as a result, the understanding of the concept shifts accordingly. These appreciations are based on Jansen's literature analysis of the comprehensive approach on an Operational, Institutional and Conceptual Level (2014) and are described as *Believers*, *Instrumentalists* and *Radical Critics*, each with a different appreciation of human security.

Believers

Believers claim that human security offers a feasible replacement for the realist view of state-centered security. Chandler illustrates this by stating that 'human security –thinking challenges the traditional agendas of power and place the needs of the individual, not states, at the center of security discourses' (Chandler, 2012, p. 224).

Tara McCormack (2011, p. 235) describes this interpretation as presenting a radical challenge to the security paradigm in which the primary concern is on the security of the state. 'In its traditional form state security is centered on the preservation of the sovereign state from external threats and activities of other states' (Chinkin, 2005, p. 1) both in terms of territory as well as in terms of 'political and physical integrity'.

Believers claim that human security can and should be understood as a realistic alternative paradigm, in which the security (and rights) of individuals can be protected, either in the broad or the narrow meaning. *Believers* would see it as a task to achieve a (sustainable) people-centered security, in which a to be defined, or contextualized balance between security and development would exist. This type of operationalization of human security can be recognized in the policies of the UNDP and some of the early adopting countries, such as Canada. The UNDP, for example, puts more emphasis on the developmental element (broad interpretation), while Canada increasingly focuses on violent threats, thus conceiving of human security in its narrower interpretation (Glasius, 2008, p. 42). Both sides would agree that the actual implementation of human security is possible and beneficial. The *believers,* as we identify them, would hold that it is possible to work from a human security concept towards a better balance between security and development strategies. This would lead to a high degree of cooperation between civil, military and local actors.

Taking the current state of the world into account, *believers* will have to admit that human security has ceased to be of importance. While political leaders from all continents are withdrawing to their national agendas, it is not possible to maintain that human security offers a feasible replacement for state security at this moment in time. As such, seen from this perspective, the comprehensive approach will likely soon disappear from international policy agendas and once again be replaced by a security-dominated paradigm.

The Radical Critics

The second category can be labeled the *radical critics*, following McCormack. She claims that human security is 'symptomatic of a disengagement of the more powerful nations from the developing world and represents the end of attempts to substantively alter non-Western societies (…) Thus the human security label is something that can be stuck on to any number of initiatives or projects, giving an appearance of strategic purpose and coherence but with little content' (2011, p. 256). The *radical critics* argue that the concept of human security aims to re-establish old or existing power relations, without taking the individual as the focal point for any strategy at all. According to this strand, the local population, instead of being a central factor, is often framed as a victim or a bystander (Glasius, 2012). This inherently is the opposite of what 'human security' (in its broad meaning: as a potentially empowering participative concept) should be aiming for, according to this interpretation.

According to *radical critics,* human security is a transformed version of state-centered security, and cannot be taken as a serious, feasible or even instrumental option for security strategies. Some *radical critics* tend to understand human security in a broad sense and do not see states capable of protecting against the insecurity that states have often created themselves. Other *radical critics* argue that using human security is just making it 'fit in the current practices' (Glasius, 2008; McCormack, 2011). *Radical critics* would dismiss the concept of human security as a basis for either doctrinal or operational strategies altogether.

According to this perspective, using a concept like 'human security' amounts to using dangerous rhetoric. Radical critics will be less surprised by the current course of events. Initiatives to create a human security force would be no option for the *radical critics*. In other words, they 'see human security less as a policy agenda within the existing structures, but rather as a radical critique of those practices' (Bellamy & McDonald, 2002, cited by Glasius, 2008, p. 39). Creating a human security force would thus, in this interpretation, either be no option, or just (dangerous) rhetoric.

Instrumentalists

Rather than seeing human security as an alternative to more traditional state-centered approaches, *instrumentalists* see human security as a welcome addition to state security. They do not interpret the shift of focus from the state to the individual as a paradigm change. Some *instrumentalists* will have a more pragmatic orientation, and understand the two types (explicitly not paradigms) of security as complementary (United Nations, 2012). Paris acknowledges the instrumentalists by stating that 'as a unifying concept for this [broad] coalition, human security is powerful', referring mainly to its political possibilities, not to its potential for academic analysis (2001, pp. 88–89). *Instrumentalists* use human security as a framework to achieve other goals—not necessarily only human security. According to an instrumentalist perspective, the concept of human security can legitimize or even give importance to certain civil and military actions. An example of this interpretation could be the 'Winning Hearts and Minds' strategy—where developmental or humanitarian activities are, for example, used to develop better local contacts. The Barcelona report provides another good example of an Instrumentalist interpretation, arguing that 'human security is vitally connected to the security of Europeans, and that the European Union therefore has a critical interest in developing capabilities to make a contribution to global human security' (Study Group on Europe's Security Capabilities, 2004, p. 28). *Instrumentalists* would make a plea for what can be framed as a 'Winning Hearts and Minds' strategy. Here the concept of human security will complement state-centered security strategies, entailing more international and inter-organizational cooperation, rather than 'true' cooperation with the people in need of 'human security'. They can be placed in between the more extreme positions of believing and radically critiquing. For them the comprehensive approach can continue to be an influential strategy. On a more critical note, this could be called 'window dressing', i.e. referring to human security while working on own (national) security.

The Future of the Comprehensive Approach—An Evaluative Perspective

The human security concept implies a significant shift in international politics from a state-centered emphasis to an emphasis on the individual and this shift indicates a different kind of commitment within international politics. The recent events in Afghanistan have shown that continuing with variations of the comprehensive approach as the main strategy for international interventions is not an obvious path to take. We argue that the future of the comprehensive approach as a feasible option for military intervention, depends on which underlying focus on human security comes to dominate policy making. Depending on which appreciation becomes dominant, the comprehensive approach can continue to play an important role.

A comprehensive approach cannot be meaningfully employed if the relevant actors (ranging from headquarter-level to boots-on-the ground level) lack a common understanding of method, goal and intention. Until today, there has been no such common understanding. Experiences with nation-building programs in the last decades have led to critique on the very possibilities of achieving such ideals. This has led to analysis of the limitations of nation-building and peacebuilding (Chandler, 2017; Lake, 2016), which essentially come down to a critique on the possibility of 'social engineering', i.e. the development of blueprint solutions to social problems (Ellerman, 2006). Over the last years, this has led to a less ambitious international agenda, an agenda more focused on domestic goals and—if abroad—on short military interventions, rather than on long-term missions. This critique on the limitations of nation-building ambitions has been explicated by the United Nations itself in a critique of its own former policy (United Nations, 2015):

> Countries emerging from conflict are not blank pages and their people are not "projects". They are the main agents of peace. However, the international approach is often based on generic models that ignore national realities.

Although these discussions about how to achieve societal transformations are certainly relevant, we argue that beneath methodological discussions about too much or too little social engineering, there is a normative discussion. In that sense, we side with Verweij, who elsewhere in this book (chapter "'Moresfare' and the Resilience Paradox: Ethics as the Terra Incognita of Hybrid Warfare and Its Challenges") refers to moresfare as 'the use or misuse of values and norms as a weapon of war or as a way to achieve a political objective'. Moresfare would be, for example, misusing values as a pretext for intervention, while the underlying intentions are political or strategic. While this position appears to be close to the radical critics discussed above, this is not so. In the same chapter, Verweij points to the relevance of Kaldor's concept of the security force and as such she does not dismiss concepts such as human security as the radical critics do. In other words, she points to the importance of taking ethical reflection seriously and asserts that such reflection should be the very foundation of intervention.

Conclusion and Discussion

In essence, the three perspectives on human security discussed above are three normative perspectives on international intervention. In the Netherlands and elsewhere, the so-called *war on terror* has been the theory underlying foreign policy from the beginning of this millennium. In a sense, this provided the opportunity for a coalition between *instrumentalists* and *believers*: interventions that followed upon realists' estimations of international situations offered the conditions for *believers* to aim at establishing human security ideals. Perhaps such a coalition can be seen as a rationale for the security-development nexus. From the perspective of *radical critics*, such a coalition is essentially a perverse coalition that a methodological discussion might even obscure. After all, idealism may constitute an outward justification for a more cynical instrumental agenda.

As regards foreign policy, while there is a broad consensus that the era of the war on terror is over, it is not clear what will follow. Some argue that we are on the brink of an era of great power politics (Biscop, 2021). What that means for the foreign policy of the EU and the Netherlands seems as yet unclear. As a firm *believer*—at least in its manifest rhetoric—in the past, the foreign policy of the Netherlands has committed to a comprehensive approach. However, it seems clear that the era of large-scale and enduring military intervention connected to a nation-building agenda is over. What will that mean for foreign policy in The Netherlands? What will that mean for the security-development nexus, and the relationship between the Ministry of Foreign Affairs and the Ministry of Defence? After all, this nexus meant that both ministries, each representing a part of the security-development equation, became connected in a specific way. If there are future military interventions, will they be focused on more limited and more achievable goals? Or will both ministries shift attention to great power politics and—for example—the different military organization that this requires? Will that lead to a totally different kind of cooperation between the Ministry of Foreign Affairs and the Ministry of Defence, one that does not rely on the security-development nexus? These are questions for further analysis research. What this chapter aimed to indicate is that answers to such questions depend on pragmatic considerations regarding achievable goals, methodological perspectives on how to pragmatically achieve such goals, but also on normative orientations.

References

AIV. (2019). *Sustainable development goals and human rights: An indivisible bond*. Advisory Council on International Affairs.

Annan, K. A. (2005). In *Larger freedom: Towards development, security and human rights for all*. United Nations Publication.

Biscop, S. (2021). *Grand strategy in 10 words: A guide to great power politics in the 21st century*. Policy Press.

Briscoe, I., & Van Ginkel, B. (2013, March). *The nexus between development and security: Searching for common ground in countering terrorism.* ICCT Policy Brief. International Centre for Counter-Terrorism.
Chandler, D. (2001). The people-centred approach to peace operations: The new UN agenda. *International Peacekeeping, 8*(1), 1–19.
Chandler, D. (2012). Resilience and human security: The post-interventionist paradigm. *Security Dialogue, 43*(3), 213–229.
Chandler, D. (2017). *Peacebuilding: The twenty years' crisis (1997–2017).* Palgrave Macmillan.
Chinkin, C. (2005). *Human security vs. state security.* Sixth session/global security environment and future alternatives. World Tribunal of Iraq Istanbul.
Clinton, H. R. (2009). *Secretary of state, testimony before the senate appropriations committee.*
Commission on Human Security. (2003). *Human security now.* United Nations.
De Coning, C., & Friis, K. (2011). Coherence and coordination: The limits of the comprehensive approach. *Journal of International Peacekeeping, 15*(1–2), 243–272.
Ellerman, D. (2006). *Helping people help themselves: From the world bank to an alternative of development assistance.* The University of Michigan Press.
Frerks, G., & Klein Goldewijk, B. (2007). *Human security and international insecurity.* Wageningen Academic Publishers.
Gabrielse, R. (2007). A 3D approach to security and development. *The Quarterly Journal, 6*(2), 67–73.
Gammer, N. (2013). Integrating civilian-military operations: The comprehensive approach and the ATF experience, 2008–2009. *Canadian Foreign Policy Journal, 19*(2), 211–222.
Glasius, M. (2008). Human security from paradigm shift to operationalization: Job description for a human security worker. *Security Dialogue Security Dialogue, 39*(1), 31–54.
Glasius, M. (2012). Human security: A shifting and bridging concept that can be operationalised. In I. Boerefijn, L. Henderson, R. Janse, & R. Weaver (Eds.), *Human rights and conflict: Essays in honour of Bas de Gaay Fortman* (pp. 149–168). Intersentia.
Hettne, B. (2010). Development and security: Origins and future. *Security Dialogue, 41*(1), 31–52.
IRSEM. (2010). *The comprehensive approach to civil-military crisis management a critical analysis and perspective.* Institute of Strategic Research at the Ecole Militaire.
Jansen, M. M. (2014). *The double bind: An analysis of the comprehensive approach on an operational, institutional and conceptual level.* Paper presented at the past, present and future of peace research. University of Groningen.
Johnstone, I., Tortolani, B., & Gowan, R. (2005). The evolution of UN peacekeeping: Unfinished business. In *Die Friedens-Warte: Blätter für internationale Verständigung und zwischenstaatliche Organisation Die Friedenswarte, 80*(3–4), 245–261.
Kaldor, M. (2007). *New & old wars.* Stanford University Press.
Lake, D. (2016). *The statebuilder's dilemma: On the limits of foreign intervention.* Cornell University.
McCormack, T. (2011). Human security and the separation of security and development. *Conflict, Security & Development, 11*(2), 235–260.
Paris, R. (2001). Human security: Paradigm shift or hot air? *International Security, 262*, 87–102.
Paris, R. (2004). *At war's end: Building peace after civil conflict.* Cambridge University Press.
Paris, R., & Sisk, T. (2009). *The dilemmas of statebuilding: Confronting the contradictions of postwar peace operations.* Routledge.
Rietjens, S. J. H., & Bollen, M. (2008). *Managing civil-military cooperation a 24/7 joint effort for stability.* Ashgate.
Schirch, L. (2011). A 3D approach to US foreign policy. *Huftington Post.* Accessed August 22, 2013. www.huffingtonpost.com/lisa-schirch/a-3d-approach-to-us-forei_b_821270.html
Study Group on Europe's Security Capabilities. (2004). *A human security doctrine for Europe: The Barcelona report of the study group on European security capabilities.* Presented to EU High Representative for Common Foreign and Security Policy Javier Solana.

Travers, P., & Owen, T. (2007). *Peacebuilding while peacemaking: The merits of a 3D approach* (UBC Center for International Relations Security and Defense Forum Working Paper #3).
Tschirgi, N. (2005, September). *Security and development policies: Untangling the relationship.* European Association of Development Research and Training (EADI) Conference.
United Nations. (2015). *Uniting our strengths for peace: Politics, partnership and people, report of the high-level independent panel on United Nations peace operations.*
UN General Assembly. (2012). *Follow-up to General Assembly resolution 64/291 on human security* (Report of the Secretary General). United Nations.
UN Secretary General. (2012). Follow-up to General Assembly resolution 64/291 on human security. In *Follow-up to the outcome of the millennium Summit 2005.* United Nations General Assembly.
UNDP. (1994). *Human development report 1994.* Oxford University Press for the United Nations Development Programme (UNDP).
UNSC. (2000). *The Brahimi report on UN peacekeeping reform.* United Nations Security Council. http://www.un.int/usa/iofact25.htm
Van der Lijn, J. (2011). *3D 'The next generation' lessons learned from Uruzgan for future operations.* Clingendael Institute, Cordaid.

Marenne Jansen is an international oriented researcher focusing on issues of security and development. She has over 15 years of professional experience, both in the field and in academia. Dr. Jansen currently works for the Advisory Council on International Affairs. She pursued her PhD in 2019 at the Radboud University, in cooperation with TNO and the Netherlands Defence Academy. Marenne holds an MA (cum laude) in philosophy from the University of Amsterdam and an MA in international political economy from the Institute of Social Studies. Her work is published internationally.

Eric-Hans Kramer is Professor in Management and Organization Studies at the Netherlands Defence Academy. Before that, he was associate professor Human Factors and Systems Safety at the same organization. He is involved in several research projects into organizational issues the military organization is being confronted with and with the development of research and educational programs into the psychosocial dynamics in military organizations.

Some Answers to Current Challenges

Contemporary Just War Thinking and Military Education

Lonneke Peperkamp and Christian Nikolaus Braun

This chapter explores the potential of the diverse field of just war theory for military education. Contemporary just war thinking takes place before the horizon of a polarised field of study (Clark, 2017, p. 331). On the one hand, one finds what has been referred to as the "traditional" or "conventional" approach that is commonly associated with Michael Walzer.[1] This approach argues for a collectivist, state-based understanding of just war. The Walzerian just war builds on the framework of international law and seeks to provide moral arguments that are of practical relevance to political and military decision-makers. On the other hand, so-called "revisionist" just war philosophers reject the collectivist starting point and advocate an individualist perspective instead. Commonly, revisionists draw their arguments from far-fetched hypothetical cases that bear little to no resemblance with real-world scenarios. Relatedly, they are mainly interested in what they call the deep morality of war, a highly idealised account of morality that takes no interest in the type of pragmatic compromises that underpin international law and the Walzerian just war. As a result of varied and fundamental disagreements, the exchange between the two just war camps has

[1] This use of terminology is debated and requires clarification. Although revisionists tend to refer to Walzer's theory as "traditional," his theory contrasts markedly with historical just war thinking.

L. Peperkamp (✉)
Faculty of Military Sciences, Netherlands Defence Academy, Breda, The Netherlands
e-mail: L.Peperkamp@mindef.nl

C. N. Braun
Radboud University, Nijmegen, The Netherlands
e-mail: christian.braun@ru.nl

E.-H. Kramer and T. Molendijk (eds.), *Violence in Extreme Conditions*,
https://doi.org/10.1007/978-3-031-16119-3_8

been limited.[2] Moreover, whilst both traditional and revisionist just war theory are based on deontological reasoning, with consequentialist elements predominantly present in the former,[3] there is also the distinct field of military virtue ethics. Rather than rights and obligations, it deals with the character traits that are presumed to be important for members of the armed forces. Although military virtue ethics is part of just war theory broadly conceived, deontological and virtue ethical reasoning are rarely combined in the literature.

Instead of discussing the substantive norms governing war, we focus in this chapter on these different theoretical approaches within the broad field of just war theory. More specifically, we critically evaluate revisionist just war theory for the purpose of educating future military leaders. Traditional just war theory and military virtue ethics have an important place in most military curricula. Walzer's *Just and Unjust Wars* (1977/2015) remains a standard text on just war theory at many military academies.[4] Whilst *jus ad bellum* is primarily relevant to statesmen and political leaders, it is assumed that future military leaders nonetheless require an understanding of the ethical principles underlying their task, the reasons for deployment, and the arguments used in the public debate.[5] *Jus in bello* specifically addresses military leaders and relates to the ethics of the profession of arms. Military virtue ethics is also widely taught at military academies as the theoretical basis for building character, including cultivating the virtues that help military professionals perform well in the extreme circumstances of war. But how about revisionist just war theory? To date, the revisionist approach has struggled to gain traction in the curricula of military academies. Traditionalists question the practical relevance of revisionist just war thinking, relegating much of it to the academic ivory tower. This seems unsurprising, as within the field of military ethics it is assumed that when something is not "helpful in providing real-world guidance for policy-makers, military commanders and leaders, or operational decision-making," there might be a place for such theorising in the discipline of philosophy, but unless it "can be brought to bear on the professional activity of military personnel in some meaningful way, they are academic exercises of interest primarily to other academics within the same field" (Cook & Syse, 2010, p. 120).

The question is: should we dismiss revisionism for the purpose of military education or can it provide valuable new insights that escape the other approaches? Whilst

[2] To divide contemporary positions in just war theory into two "camps" is undisputedly simplistic. Nevertheless, there is an evident split between two main approaches: Walzerian just war theory, called the "traditional" or "conventional" or "orthodox" position, because it has been taken as the starting point of inquiries within this field for decades. Far from adopting a coherent position, revisionists generally set out to revise the traditional theory. For analyses of these two approaches see e.g.: Lazar (2017), Pattison (2018) and Braun (2018). For a critique of the appropriateness of the term "revisionism" see e.g. Steinhoff (2012).

[3] See Benbaji and Statman (2021) for a contractarian account of just war, and Shaw (2016) for a purely utilitarian account of just war.

[4] Which is not to say that there is no attention for other authors, or that Walzer's work is used at every military academy.

[5] Moreover, soldiers are bound to obdedience, but never cease to be members of society. Therefore, they have a say in the political debate and they are entitled to an opinion on the *jus ad bellum*.

in our work we generally see more merit in traditional just war theory, we think that the role of revisionism deserves more attention than it has received so far.[6] In order to assess that role, we seek to connect just war thinking to the German and Dutch leadership concepts: *Innere Führung* and the ideal of "thinking soldier." In the following two sections, we will discuss traditional and revisionist just war theory. Subsequently, we will provide an overview of the German and Dutch leadership concepts. We side with critics who argue that the practical applicability of revisionist just war thinking is limited. However, we will argue that whilst the space for reflective moral decision-making is very limited on the battlefield, revisionist just war thinking may play an indirect role at military academies. The benefit of revisionism, we will argue, ties into the ideal of *Innere Führung* and the "thinking soldier." At the same time, however, its shortcomings do not only underline the benefits of traditional just war theory, they also bring us back to the importance of virtue ethics and character building in military education. Knowledge of rules and principles and the ability to critically reflect on that normative framework only takes us so far. Doing "the right thing" in the extreme circumstances of war requires not only *knowing* what to do, but also requires the *motivation* to do it. In other words, the ethical competence of military leaders must reflect a cognitive and motivational dimension. Strengthening such competence benefits from a comprehensive theoretical basis; one that combines the various theoretical approaches.

Walzer's Just War

Walzer has always been sceptical that philosophical reflection alone can address the moral questions arising in war. His is a "practical morality," which does not directly engage with "the most profound questions of moral philosophy" (Walzer, 2015, p. xxii). Relatedly, Walzer advocates a case-based approach to just war. For him, it is the reality of war, as expressed in the experience soldiers gain on the battlefield, that provides the material to argue about the morality of war. An important underlying assumption of conventional just war theory is that war is an exceptional moral domain which is distinct—although not separate—from normal morality. There are several reasons to defend such "exceptionalism:" the magnitude and scale of war as a violent conflict; the fundamental political interests at stake; the uncertainty and fog of war; and the general non-compliance with fundamental moral norms such as the obligation to respect the right to life (Shue, 2008, p. 88; Walzer, 2015, pp. 335–345).

Taking the "legalist paradigm" as his starting point allows Walzer to follow the laws of war closely.[7] His is therefore a legalistic theory, and some (see, e.g. Lazar, 2017, p. 38) have argued that his main objective is to provide the laws of

[6] In previous work, we both discussed the benefits and drawbacks of these two camps in contemporary just war theory. This chapter builds on that work: Braun (2018) and Peperkamp (2019).

[7] Walzer's (2015, pp. 61–63) legalist paradigm is grounded in the assumption that "international society has a law that establishes the rights of its members—above all, the rights of territorial

war with a moral foundation.[8] Nonetheless, Walzer considers positive international law to be "radically incomplete" (2015, p. xxvi). Moreover, the "war convention" is comprehensive; it consists not only of legal regulations, but also moral norms, customs, professional guidelines, religious/philosophical principles and reciprocal understandings that shape our judgements about military action (Walzer, 2015, p. 44). It is, in other words, the agreed upon set of norms and values on which the international moral and legal order is based. Given the importance of political sovereignty and territorial integrity in that international order, it is states that are taken to be the most important actors.[9] However, Walzer states that the war convention is "necessarily imperfect [...] because it is adapted to the practice of modern war. It sets the terms of a moral condition that comes into existence only when armies of victims meet" (Walzer, 2015, p. 45). With his account of just war, Walzer created a normative common language that enables concrete judgements in the political reality. It is a practical moral theory, in line with international law, congruent with common sense morality, and providing specific guidelines regarding military behaviour in the extreme situation of war.

Those specific guidelines can be broken down into three branches: *jus ad bellum* entails norms on the justification of war, *jus in bello* norms on proper behaviour in war and *jus post bellum* norms on how to realise a just peace.[10] Central in Walzer's just war theory and in international law is the "independence thesis:" *jus ad bellum* and *jus in bello* are separate branches. *Jus ad bellum* is asymmetrical, whilst the subsequent warfare is governed by moral rules that are the same for just and unjust combatants alike. Unlike *jus ad bellum, jus in bello* is symmetrical; combatants have equal rights and obligations in war. That means that whilst non-combatants are immune and cannot be intentionally targeted, combatants are equally liable to be killed and equally permitted to kill their adversaries. This central idea is reflected in international humanitarian law, which determines that its norms apply to all those concerned and imposes the same obligations on them. Whilst Walzer starts his argument stating that he does not consider the legalist paradigm as sacrosanct (Walzer, 2015, p. xxvi), he nonetheless employs the legalist paradigm as a "frame" (Johnson, 2014, p. 5). This then, lets him accept positions that earlier just war thinkers would have rejected as morally indefensible, such as his argument for a moral equality of combatants that

integrity and political sovereignty." Thus, whilst he at times argues for exceptions to the established laws of war, he takes them as the starting point of his moral argument.

[8] It is no surprise, then, that Walzer's just war has been identified as being liable to conservatism and relativism at the same time. Regarding the former aspect, Walzer seems to remove the critical function of just war by building his moral argument on the legalist paradigm. Regarding the latter, his argument could be seen as relativist in the sense that he follows the development of the legality of war and, therefore, emphasises changes in international society over moral principles (see O'Driscoll 2008, pp. 96–98).

[9] These key elements are grounded in the Peace of Westphalia (1648) that concluded the Thirty Years' War.

[10] For Walzer's treatment of *jus post bellum*, an area of just war that he does not specifically discuss in *Just and Unjust Wars*, see Walzer (2012).

abandons the requirement to discriminate between those who fight a just war and those who fight an unjust war.

The Revisionist Critique

Revisionism is an umbrella term for a critical current within just war theory.[11] Although there are many different revisionists, they generally set out to revise the norms of traditional just war theory. Moreover, they are like-minded when it comes to the theoretical underpinnings on which their theory is based. The starting point of these theorists is not the legalist paradigm, or the reality of warfare, or historical just war theory. They take Walzer's theory as the "conventional" ruling theory that must be checked for logical incoherence with the goal of constructing a better theory. In order to do that, they rely on Rawls's method of reflective equilibrium (Rawls, 1971/1999, pp. 18–43).[12] The main objective of revisionists is to write novel philosophy on war-related issues such as the ethics of harming, the duty to save, or political authority. It seems fair to state that most revisionists do not share Walzer's ambition to provide a practical morality that can easily be taken to inform political and military decision-making. The foremost expression of their limited interest in the experiences men and women undergo on the battlefield is their use of far-fetched, oftentimes other-wordly, thought experiments.[13] Consider, for example, the following thought experiment Helen Frowe provides in the context of self-defence:

Ray Gun

> Falling Person has been blown by the wind down a well, at the bottom of which you are trapped. Falling Person will crush you to death unless you vaporise her with your ray gun. If you do not vaporise her, your body will cushion Falling Person's landing, saving her life.[14]

This methodology, i.e. the use of such thought experiments as building blocks in the reflective equilibrium, fits well into the abstract process of conceptual analysis that characterises analytical philosophy. Unsurprisingly, however, non-revisionists question the practical relevance of such hypotheticals.

Revisionists furthermore object to the important position of the state in Walzer's theory. They reject what they see as a collectivist approach to war and hold that the responsibility for killing resides in the individual. In contrast, they advocate

---

[11] At the risk of excluding prominent revisionist just war thinkers, we consider the work of Cécile Fabre, Helen Frowe, Jeff McMahan and David Rodin as characteristic of the revisionist enterprise.

[12] Rawls' reflective equilibrium is a philosophical method that constitutes a deliberative process in which a theory is developed or taken that yields certain principles and to compare those with our considered judgements, i.e. the intuitions we have in specific situations with regards to those principles. In doing so, principles and practice are balanced in an attempt to achieve such reflective equilibrium.

[13] Whilst the methodological difference marks an important aspect of their disagreement, our focus in this chapter is on the practical applicability of just war theory.

[14] Frowe (2016, p. 17).

an approach they refer to as "reductive individualism." That approach is reductive because it assumes that the rules that regulate killing in war can be reduced to those regulating interpersonal killing outside of war (Frowe, 2014, p. 13). At the heart of reductivism is the conviction that there is only one set of moral principles which applies all the time; there is nothing inherently special about war. Consequently, the idea that there are different moral domains such as war and peace is a non-starter for revisionists. At the same time, most revisionists are individualists, as they argue that moral theory must concentrate on individuals rather than collectives.

As a result of these fundamentally different underpinnings, revisionists object to key elements of Walzer's theory. Aside from aiming to revise *jus ad bellum*, many revisionists have targeted the rules of *jus in bello*, including the independence thesis, and consequently the moral equality of combatants and the immunity of non-combatants.[15] Following reductive individualism, war is simply an aggregate of just or unjust acts of individual self- and other defence on a large scale.[16] This means that, like *jus ad bellum, jus in bello* is asymmetrical: unjust combatants are responsible for posing an unjust threat and are therefore liable to attack. Consequently, just combatants are permitted to kill unjust combatants, but not vice versa, since just combatants have done nothing to make themselves liable to be killed. Put differently, in order to be subjected to force, the unjust threat must have forfeited its right not to be harmed, which revisionists tie to the just cause of self and other defence. McMahan argues that: "the principles of *jus ad bellum* apply not only to governments but also to individual soldiers (*agents*), who in general ought not to fight in wars that are unjust. It denies that *jus in bello* can be independent of *jus ad bellum* and concludes that in general it is not possible to fight in a way that is objectively permissible in an unjust war" (McMahan, 2012). There is no moral equality of combatants; unjust combatants cannot do anything right, except lay down their weapons (Lazar, 2017, p. 38). This approach, then, also challenges the principle of non-combatant immunity. When civilians contribute to an unjust threat, or when they are morally responsible for such a threat, they make themselves liable to be killed. According to the revisionist approach, some non-combatants can be legitimate targets.

In summary, the objective of revisionists is to formulate norms that are derived from abstract moral principles. Through this abstraction and the use of hypothetical examples, an ostensibly greater understanding is gained. As Frowe (2014, p. 5) puts it succinctly: "Stripping away the detail can enable us to identify general principles that can be obscured by the intricacies of historical cases." Complex reality can mask things, it is assumed, and norms can be more easily identified without the fog that surrounds historical events (Frowe, 2016, p. 1). The practical objective, however, hardly comes into focus as a result. Revisionists generally show little interest in the war conditions in which moral decisions are made. As a result, whilst their arguments may be philosophically coherent, they are also difficult to apply. Do combatants have enough knowledge of political dynamics to judge whether their war is just? Is that knowledge available to the general public at all? Are they able to determine which

[15] See further e.g. Lazar (2018).

[16] E.g. Rodin (2003).

individuals amongst the combatants and civilians are liable to attack, and who are not? And in the extreme conditions of war—characterised by stress, agony, time pressure, limited opportunities for rational decision-making—can combatants make such judgements? The time for reflective moral decision-making on the battlefield is very limited indeed. In war, therefore, the guidance provided by revisionist theory can hardly be followed.

Therefore, unsurprisingly perhaps, revisionist just war thinking has not featured in any prominent way on the curricula of military academies. The reasons seem obvious: military academies educate and train future military practitioners. The focus is on providing action-guidance, not on creating philosophers. In addition, as will be discussed in the next section, the revisionist challenge to Walzer's theory comes with the danger of undermining the laws of war. Relatedly, the individualist understanding of revisionists is in tension with the collectivist and hierarchical understanding that undergirds the organisation of the military profession.[17] These weighty reasons notwithstanding, we will argue in the following that there is a limited place for revisionist just war thinking in military academies.[18] As we will demonstrate, that place cannot be where soldiers are being prepared for practical moral decision-making on the battlefield. Rather, revisionist just war thinking could play a role in the endeavour to produce the type of military leaders imagined by the German and Dutch leadership philosophies.

Military Leadership Philosophies

What is the relevance of revisionist just war theory for the purpose of military education? Before we can properly explore the educational aspect of just war thinking, and the role of revisionist just war theory within it, we need to turn to the German concept of *Innere Führung* (leadership development and civic education) and the Dutch concept of the "thinking soldier."[19] Once we have established the relevance of these conceptualisations, we will be in the position to determine the limited place for revisionist just war thinking in military education. The German and Dutch leadership philosophies are related; we identify interesting parallels between the German *Innere Führung* and the Dutch conceptualisation. Both share the objective of educating self-reflective and politically sensitive soldiers who act responsibly and take responsibility for their actions. Seeing the ideal military leader as a critical thinker—an individual

[17] See further on these and other dangers: van Baarda (2015).

[18] Interestingly, at the 2021 McCain Conference, Edward T. Barrett, Research Director at the U.S. Naval Academy's Stockdale Center for Ethical Leadership, noted that, at present, their core ethics course concentrates on the "traditional" perspective as manifested in Walzer and he identified a need to also consider the revisionist just war.

[19] It seems that the German conceptualisation is more far-reaching than the Dutch, as the Bundeswehr's *Innere Führung* is meant to be understood as a "concept of reform" that seeks to distance itself from the leadership philosophy that undergirded Hitler's Wehrmacht. See Bundesministerium der Verteidigung (2008, § 205).

who is academically trained, aware of his/her own moral norms and values, and sensitive to his/her place in the social and political context—can help us locate a modest slot for revisionist just war thinking.

Seeking to learn from Germany's militaristic past, the Bundeswehr is supposed to be anchored at "the centre of society" (Bundesministerium der Verteidigung, 2008, § 101). Its soldiers are considered to discharge their responsibilities when they "out of inner conviction, actively stand up for human rights, freedom, peace, justice, equality, solidarity and democracy as the guiding values of our state" (Bundesministerium der Verteidigung, 2008, § 106). Command, obedience and ethics are the key principles of this concept and reflect the "remarkable trinity of political purpose, constitutional essentials, and soldierly command and discipline in the blast of fighting, anger, and hatred native to war and political violence in its variety" (Abenheim & Halladay, 2016, p. 3). One element of this understanding is the requirement that soldiers "think for themselves, rather than obey blindly" (Bundeswehr, 2021). Some (e.g. Hartmann, 2016, p. 23) argue that it is in this "primacy of conscience" where the German leadership philosophy differs from that of some of its allies. In a nutshell, the idea is to foster an environment where *Innere Führung* is internalised; the thinking soldier, who actively grapples with the difficult decisions he/she may be required to take, becomes the ideal soldier: "Innere Führung forms the spiritual and moral basis of the armed forces" and "it enables *acting upon insight* (emphasis added)" (Bundesministerium der Verteidigung, 2008, § 107). That said, as Peter Olsthoorn (2016, p. 35) notes, it is important to point out that *Innere Führung* is not meant to resemble simply "an internal moral compass" in which societal values play no role: "One of the ideas behind *Innere Führung* is that soldiers should disobey orders that are manifestly unethical; they are expected to think for themselves. But that does not mean that anything goes; the values that should guide that independent thinking are clearly societal."

As the previous citation invokes the concept of a thinking soldier, it acts as a bridge to the Dutch leadership philosophy. Central in the education of the Netherlands armed forces is the ideal of the thinking soldier.[20] It assumes that the intellectual and pragmatic challenges of today's complex security environment make it essential for "officers to keep a broad focus and an open mind and, most importantly, to take responsibility for their actions, as these do not only affect their own lives, but the lives and livelihoods of others as well" (Oonincx, 2019, p. xi). The complexity of the military profession, in other words, requires a combination of academic education and a more practical preparation for the military profession (Oonincx, 2019, p. vi). Officers need both academic skills *and* the ability to act effectively in practice. The concept of the thinking soldier builds on Donald Schön's idea of a "reflective practitioner;" an academic practitioner who continuously questions, reflects and reforms assumptions, and in this way, is aware of his/her (implicit) knowledge and learns from experience (Schön, 1992). Such a practitioner comes up with creative practical solutions on the basis of academic reflection and analysis, in complex situations where a simple application of rules or academic knowledge will not do the trick.

[20] See van Baarda and Verweij (2006).

The concept of the "thinking soldier" reflects the idea that military leaders must be scholars *and* soldiers; they combine both the "habitus of the scholar," i.e. critical thinking, analysis, systematic doubt and the "habitus of the officer," i.e. authority, agility and the ability to command (including swift decision making, analysis with limited information, pragmatism) (Jansen et al., 2019, pp. 340–341). In that way, "a reflective practitioner is a craftsman thinking about his job by asking questions" (Bijlsma, 2019, p. 117). Ger van Doorn (2019, pp. 162–163) subsequently draws specific attention to ethical reflection. He sees a reflective practitioner as someone who constantly reflects on the norms and values that ground our behaviour, and who is able to transform that into (self-) insightful action in daily practice. It is the combination of ethical reflection and operational action that creates thinking soldiers (Van Doorn, 2019, p. 178).[21]

Critical Thinking and Revisionism

Having laid out two specific contemporary military leadership philosophies, does that indicate a place for revisionist just war thinking? As stated in the introduction, given the practical orientation of military education, traditional just war theory has been taught predominantly at military academies. Nonetheless, we believe that the revisionist just war can play a helpful role in educating military decision-makers, as it connects to the rationale behind *Innere Führung* and the thinking soldier. Returning to the disagreement between Walzerians and revisionists regarding the moral equality of combatants and relatedly the relationship between the morality and laws of war, we have seen that revisionists, much more empathically than Walzerians, demand that combatants engage with the *jus ad bellum* decision, not just with *jus in bello* decisions. In fairness to Walzer, he does not entirely disconnect *jus ad bellum* and *jus in bello* decisions. Discussing the Second World War on the German side, which for Walzer was the prototypical unjust war, he argues (2015, p. 345) that the best moral option would have been for Wehrmacht soldiers not to participate in it: "...soldiers have a right to refuse to fight in a war they believe to be unjust; ... But it is an act of heroism, and it can't be morally required; unheroic conduct isn't criminal conduct." In a sense, one might add, Walzer is again suggesting a pragmatic compromise here. Many soldiers do not have the information or the education to judge the justifiability of the war they are called to fight. However, Walzer clearly accepts exceptions to what might be called a presumption to participate.

Revisionists, of course, go much further than Walzer regarding the responsibility they allocate to the individual and the decision to participate in war. This might just be what *Innere Führung* and the thinking soldier require. As Bernhard Koch (2019, p. 7) notes, in the eyes of revisionists, soldiers "must therefore ensure they are aware of the ethical reasons for their deployment, and cannot simply shrug off their moral responsibility as a question of command and obedience." Whilst the

[21] Verweij (2020).

jus ad bellum decision remains a political decision, we think that having future military leaders engage with the moral arguments of revisionists, some of which challenge the "traditional" account of just war and the established laws of war, can create more conscientious practitioners. It allows a deeper understanding of the normative framework and fosters the political and social sensitivity that is central in these leadership philosophies. In that sense, revisionist just war thinking can help strengthen ethical competence.

To see how this could work, let us take a closer look at what ethical competence entails. Gerhard Kruip (2019) makes a helpful distinction between two levels of ethical competence.[22] Firstly, there is a cognitive, argumentative dimension. That enables officers to test the moral purchase of norms and to judge the morality of specific situations. However, the cognitive, argumentative dimension marks only a partial competence and needs to be combined with the emotional and motivational dimension. The latter aspect relies on internalisation based on examples, encouragement and recognition: "That is, if we wish to derive individual conclusions for action from moral norms, we have to apply those norms to specific situations and also analyse these situations as best we can" (Kruip, 2019, pp. 12–13). Revisionist just war thinking could be employed to analyse the moral purchase of Walzer's traditional just war. Pointing to the chasm between the ideal-type morality revisionists call the deep morality of war and the established laws of war that form the bedrock of Walzer's theory, would enable military decision-makers to better understand the uneasy compromises that have resulted in today's laws of war—and why they need to be upheld.[23] Whilst discussion may lead to the conclusion that revisionists are right that just and unjust combatants do not face each other as *moral* equals, it would also point to the weighty moral reasons for granting equal rights to combatants on both sides. In other words, revisionists reveal both weakness and strength of the Walzerian theory. Whilst Walzerians might be wrong in the ideal, granting equal rights to combatants on both sides is a pragmatic compromise that is instrumental for the constraint of war. Therefore, equipping military leaders with a grasp of where this compromise originates, and why it matters, will help them put it in action. In particular, awareness would be raised that an asymmetric application of the laws of war would potentially have dramatic consequences vis-à-vis how wars should be conducted.[24]

As David Rodin and Henry Shue (2008, p. 7) note, "it is likely that most combatants would, rightly or wrongly, view themselves as fighting for the just side and would therefore attribute to themselves any asymmetric war rights reserved for the just side. In this way, the overall destructiveness of war would go up with no strategic advantage being reaped by the genuinely just side." Engaging with the revisionist

[22] Cf. the distinction between *Ausbildung* and *Bildung*. See e.g. van Baarda and Verweij (2006, pp. 31–33).

[23] For an overview of just war and the laws of war as historically conditioned realities, see Johnson (2017).

[24] See also Mavrodes (1975) for an argument as to why traditional just war theory might not reflect any deep moral truth, it is a useful convention for limiting these dramatic consequences of war.

critique could point to the conceptualisation of the laws of war as a non-ideal compromise that has been accepted internationally and, therefore, functions as an important mechanism of restraint in war. It could thus inform soldiers about the need for a pragmatic objection against aligning the "deep morality of war" with the laws of war, which even McMahan himself accepts.[25]

It goes without saying that the approach of scrutinising and consequently reaffirming the established laws of war is intellectually demanding of military leaders. It is also an undertaking that needs to be facilitated carefully in order not to risk the fragile construct of the laws of war. Jeremy Waldron, for example, cautions moral philosophers to "take special care" when evaluating and challenging the laws of war: "These laws are not robust, they are not particularly resilient, they are difficult to enforce, and they depend largely on the voluntary self-application of problematic and constable norms to the conduct of groups of men who find themselves in circumstances of mortal danger" (Waldron, 2018, p. 81). The fear, that is, is that whilst revisionist just war thinkers may be able to undermine the established laws of war, they are unable to put in place laws that align with the morality of war more closely.

We share Waldron's concern about the potential negative repercussions of revisionist just war thinking for the conduct of war. An ostensibly easy solution would be to simply ignore it in military education: "These are the laws of war and you must not question them." In contrast, we think that we can do better. In line with the concepts of *Innere Führung* and the thinking soldier as we understand it, we propose to engage and reflect on those laws with future military leaders. Rather than presenting the established laws of war as a robust and resilient body of law that has no moral alternative, we propose to discuss it openly as the perhaps uneasy but necessary moral compromise that it is. With the help of the revisionist argument, officers would be able to better understand its moral foundations *and* its shortcomings, but also internalise why it must not be undermined. In that way, we believe, "thinking soldiers" would uphold the laws of war by *acting upon insight*, rather than simply obeying what they are told. As a result, the revisionist challenge to Walzer's traditional just war, and the laws of war on which it is built, can make a valuable contribution to our military academies.

A Comprehensive View on Just War Theory

Whilst we have argued for a modest role for revisionist just war thinking, it should come as no surprise that we consider it to have limited value for other crucial elements

---

[25] In fact, McMahan (2006, p. 38) is happy to acknowledge that arguing about the morality and legality of war are different undertakings: "It is important to understand that the account I have developed of the deep morality of war is *not* an account of the laws of war. The formulation of the laws of war is a wholly different task, one that I have not attempted and that has to be carried out with a view to the consequences of the adoption and enforcement of the laws or conventions. It is, indeed, entirely clear that the laws of war must diverge significantly from the deep morality of war as I have presented it."

of military education, especially with regards to preparing soldiers for the moral and legal questions that await them on the battlefield. Revisionist methodology, especially the individualist underpinnings and the use of other-worldly thought experiments, removes them from the reality of war. It seems questionable that this can helpfully inform military decision-making. In that sense, we are sympathetic to Walzer's engagement with the experiences men and women gain on the battlefield as a tool to derive norms but also test moral judgements. Like Walzer's conceptualisation of just war, ours seeks to be a realistic one that is capable of providing action-guidance. Such an account departs from the collectivist nature of warfare, the institutional context and the social practices related to it. Our concern is that, because of not translating the proposed norms to the context of war, revisionist just war theory loses the "close linkage of decision-making and concrete action" (Reichberg, 2018a, p. 65) that was a central aspect to earlier modes of just war thinking and that remains of crucial importance for practitioners today.

Moreover, and returning to Kruip's distinction of two levels of ethical competence, revisionists do not aim to capture the emotional and motivational level. It is here where we think that training in the military virtues is especially important, complementing knowledge of the laws of war and traditional just war theory: it strengthens the emotional and motivational dimension of moral competence that escapes revisionists.[26] Especially given the complex security environment, military leaders will be confronted with moral dilemmas, a lack of clear legal answers, or inclinations *not* to do the "right thing." David Perry (2016, p. 5) rightly notes that "ethical decision-making cannot be reduced to a short checklist or model"; it requires a wide range of moral emotions. No checklist, model, or set of norms can provide answers in every situation. In the same vein, Désirée Verweij and Tine Molendijk show that such a check-list view, separating the various just war criteria and cutting loose the justification of the use of force from the justification of warfare, risks losing sight of the context in which force is used and specifically the purpose of the endeavour itself, i.e. correcting the wrong that was the cause for war (Verweij and Molendijk, 2019).[27] Therefore, we think that a thorough understanding of the laws of war and their moral foundations as explicated in just war theory, in addition to a strong moral compass and cultivated virtues, best supports the decision-making process in challenging situations.[28] As stated in the introduction, however, the various theoretical approaches within the broad field of just war theory are rarely combined.

Interestingly, as an engagement with the just war tradition demonstrates, thinkers as early as Thomas Aquinas in the thirteenth century, who himself drew on the

[26] See also Verweij (2020, pp. 19–20).

[27] And that view of the context and purpose, they furthermore argue, helps to protect combatants from moral injury.

[28] Along those lines, see David Perry (2016, p. 4): "We need wisdom and critical thinking as well as a good conscience.".

thinking of Aristotle, emphasised the importance of virtue for the military profession.[29] Aquinas was well aware of the fact that soldiers need virtues in order to deal with the emotions and passions that inevitably arise in mortal combat. Aquinas's just war, as he let it unfold within an account of the virtues, recognises that there is no time for imaginative modelling in the heat of battle there: "Aquinas's virtue approach has, by contrast, the advantage that it is designed specifically for such settings; thus instead of first separating reflection from practice and then facing the challenge of reuniting the former with the latter, Aquinas attempts a unified account that joins the two from the beginning."[30] Fast forwarding to the contemporary era, it is interesting to note that the revival of just war theory that Walzer triggered with his *Just and Unjust Wars* coincided with a resurgence of virtue scholarship. However, as Michael Skerker notes (2019, p. xxv), the two revivals did not immediately lead to a "profusion of texts about *military* virtue."

Neither are these two approaches, just war theory and military virtues, often merged together. There are exceptions, though, and we think that such comprehensive accounts of just war theory are especially valuable. Allen Buchanan, for example, assumes that the goal of just war theory is not limited to listing theoretical justifications (such a "checklist" of criteria), but also includes providing "directly action-guiding rules," guidance for the evaluation of institutional processes, criteria for the evaluation of the laws of war, the decisions of leaders, and social practices, plus an account of the virtues of leaders.[31] Another exception is A. J. Coates (2016, pp. 1–19), whose secular account draws on the classical idea of *bellum justum*. He argues that the key determinants of justice in war are the moral dispositions of combatants. It is not only about rules and principles, but also virtues and vices. Since virtues and vices are expressions of combatants' moral character, they therefore incline or disincline towards moral or immoral conduct. Coates (2016, p. 15) also emphasises that moral agency is both cognitive and volitional. Consequently, even if someone knows the correct action to take, that does not necessarily mean that this person will act accordingly. Therefore, the moral agent needs correctly-ordered virtues in order to "will" the right action.

Concluding Thoughts

We have argued that there is a modest place for revisionist just war thinking in the education of military decision-makers. Engaging with the revisionist challenge to traditional just war theory facilitates a deep understanding of the laws of war and an awareness of the necessity to uphold them, which is based on insight, not just

[29] See Aquinas, *ST*, II-II, q. 40, a. 1. Reichberg (2018b, p. 53) argues that Aquinas's emphasis of virtue (regarding war, especially the virtues of military prudence and battlefield courage) can be seen as his most important contribution to just war theorising.

[30] See Reichberg (2018b).

[31] Note that Buchanan lists more goals, see further Buchanan (2018, pp. 69–71).

on obedience. Nonetheless, we have also pointed out a number of drawbacks of revisionist just war theory: the risk of undermining the laws of war, the inability to provide action-guidance (since rules are derived from abstract moral principles and hypothetical thought experiments), and the neglect of the internal motivation crucial on the battlefield. There is some, albeit limited, value in integrating revisionism in military curricula.

For the purpose of military education, we especially see merit in a comprehensive approach (traditional just war theory with reference to the revisionist critique) that joins together reflection and practice. Such an approach is likely to be most effective at preparing military practitioners to bear what Martin Cook (2006, p. 27) refers to as the officer's "weight of responsibility," namely, to "thoroughly incorporate thought about the *jus in bello* side of just war into standard operating procedure." And whilst perhaps less obvious, the ability to reflect on political *jus ad bellum* issues further strengthens the ethical competence of military decision-makers, as it deepens insight into the nature of their task and enables them to position themselves in the wider society.[32] As Jansen and Verweij (2019, pp. 61–62) put it, the challenges of today's security environment require "a certain level of self-realisation and individual development in dialogue with the wider (globalised) world." Such reflection on the legal and moral rules strengthens the internal motivation to act on them, and as such, helps to increase compliance with the laws of war. Importantly then, this must be complemented by a virtue ethics approach, which is essential for further strengthening the second dimension of moral competence and helps to internalise the moral and legal rules. Given the extreme circumstances and the confrontation with violence, military leaders need the strength of character to do the right thing in complex situations, despite risks, and given the multiple demands and values at stake. When it comes to the value of just war theory for military education, we think that a practically-oriented and unified approach, one that draws on reflection and virtue, is most helpful in achieving the ideals of *Innere Führung* and the "thinking soldier."

References

Abenheim, A., & Halladay, C. C. (2016). Professional soldiers and citizens in uniform: Some thoughts on *Innere Führung* from a transatlantic perspective. *Ethics and Armed Forces, 1*, 3–7.

Aquinas, T. (1948). *Summa Theologica*, translated by the Fathers of the English Dominican Province. Allen, TX: Christian Classics.

Benbaji, Y., & Statman, D. (2021). *War by agreement: A contractarian ethics of war*. Oxford University Press.

Bijlsma, T. (2019). From thinking soldiers to reflecting officers—Facts and reflections on officers' education. In W. Klinkert, M. Bollen, M. Jansen, H. de Jong, E. H. Kramer, & L. Vos (Eds.),

[32] George Lucas also mentions the responsibility of senior military leaders to provide professional advice related to political *ad bellum* decisions, "in order to protect their profession's future leaders from being forced to commit professional malfeasance by fighting in wars of questionable moral justification." Lucas (2016, p. 115).

NL ARMS Netherlands (annual review of military studies 2019) educating officers: The thinking soldier—The NLDA and the Bologna declaration (pp. 115-131). TMC Asser Press.
Braun, C. N. (2018). The historical approach and the "war of ethics within the ethics of war." *Journal of International Political Theory, 14*(3), 349–366.
Buchanan, A. (2018). *Institutionalizing the just war*. Oxford University Press.
Bundesministerium der Verteidigung. (2008). *Innere Führung: Selbstverständnis und Führungskultur*. https://www.bmvg.de/resource/blob/14258/a0e22992bc053f873e402c8aaf2efa88/b-01-02-02-download-data.pdf. Accessed 16 October 2021.
Bundeswehr. (2021). *Die Innere Führung – Das Wertegerüst der Bundeswehr*. https://www.bundeswehr.de/de/ueber-die-bundeswehr/selbstverstaendnis-bundeswehr/innere-fuehrung. Accessed 21 October 2021.
Clark, I. (2017). Taking 'justness' seriously in just war: Who are the 'miserable comforters' now? *International Affairs, 93*(2), 327–341.
Coates, A. J. (2016). *The ethics of war*. Manchester University Press.
Cook, M. (2006). Ethical Issues in war: An overview. In J. B. Bartholomees (Ed.), *U.S. Army War College guide to national security policy and strategy*. Strategic Studies Institute, US Army War College.
Cook, M. L., & Syse, H. (2010). What should we mean by 'military ethics'? *Journal of Military Ethics, 9*(2), 119–122.
Frowe, H. (2014). *Defensive killing*. Oxford University Press.
Frowe, H. (2016). *The ethics of war and peace: An introduction*. Routledge.
Hartmann, U. (2016). What's the matter with *Innere Führung*? *Ethics and Armed Forces, 1*, 22–26.
Johnson, J. T. (2014). *Sovereignty: Moral and historical perspectives*. Georgetown University Press.
Johnson, J. T. (2017). A practically informed morality of war: Just war, international law, and a changing world order. *Ethics & International Affairs, 31*(4), 453–465.
Koch, B. (2019). The relationship between peace ethics, military ethics and security policy. *Ethics and Armed Forces, 2*, 4–10.
Kruip, G. (2019). Values and norms: Don't 'teach', encourage independent acquisition! *Ethics and Armed Forces, 2*, 11–15.
Lazar, S. (2017). Just war theory: Revisionists versus traditionalists. *Annual Review of Political Science, 20*, 37–54.
Lazar, S. (2018). Method in the morality of war. In H. Frowe & S. Lazar (Eds.), *The Oxford handbook of ethics of war* (pp. 21–40). Oxford University Press.
Lucas, G. R. (2016). *Military ethics: What everyone needs to know*. Oxford University Press.
Jansen, M., Braender, M., & Moelker, R. (2019). What sets the officer apart? Dutch and Danish educational reforms leading to the habitus of the thinking soldier. In W. Klinkert, M. Bollen, M. Jansen, H. de Jong, E.H. Kramer, L. Vos (Eds.), *NL ARMS Netherlands (annual review of military studies 2019) educating officers: The thinking soldier-The NLDA and the Bologna declaration* (pp. 337–353). TMC Asser Press.
Jansen, M., & Verweij, D. (2019). Officer Bildung: A philosophical investigation of preconditions for military character education. In W. Klinkert, M. Bollen, M. Jansen, H. de Jong, E. H. Kramer, & L. Vos (Eds.), *NL ARMS Netherlands (annual review of military studies 2019) educating officers: The thinking soldier—The NLDA and the Bologna declaration* (pp. 59–75). TMC Asser Press.
Mavrodes, G. I. (1975). Conventions and the morality of war. *Philosophy & Public Affairs, 4*(2), 117–131.
McMahan, J. (2005). Just cause for war. *Ethics & International Affairs, 19*(3), 1–21.
McMahan, J. (2006). The ethics of killing in war. *Philosophia, 34*(1), 23–41.
McMahan, J. (2012, November 12). Rethinking the 'just war,' Part 2. *The New York Times*. https://opinionator.blogs.nytimes.com/2012/11/11/rethinking-the-just-war-part-2/. Accessed 18 March 2022.
O'Driscoll, C. (2008). *Renegotiation of the just war tradition and the right to war in the twenty-first century*. Palgrave Macmillan.

Olsthoorn, P. (2016). Integrity, moral courage and *Innere Führung. Ethics and Armed Forces, 1*, 32–36.
Oonincx, P. (2019). Foreword. In W. Klinkert, M. Bollen, M. Jansen, H. de Jong, E. H. Kramer, & L. Vos (Eds.), *NL ARMS Netherlands (annual review of military studies 2019) educating officers: The thinking soldier—The NLDA and the Bologna declaration* (pp. v–vi). TMC Asser Press.
Pattison, J. (2018). The case for the nonideal morality of just war: Beyond revisionism versus traditionalism in just war theory. *Political Theory, 46*(2), 242–268.
Peperkamp, L. (2019). De Oorlog in de Theorie van de Rechtvaardige Oorlog. *Algemeen Nederlands Tijdschrift voor Wijsbegeerte, 111*(1), 63–94.
Perry, D. L. (2016). *Partly cloudy: Ethics in war, Espionage, covert action, and interrogation.* Rowman & Littlefield.
Rawls, J. (1999). *A theory of justice.* Harvard University Press.
Reichberg, G. M. (2018a). Historiography of just war theory. In H. Frowe & S. Lazar (Eds.), *The Oxford handbook of ethics of war* (pp. 59–79). Oxford University Press.
Reichberg, G. M. (2018b). Thomas Aquinas (1224/5-1274). In D. R. Brunstetter & C. O'Driscoll (Eds.), *Just war thinkers: From Cicero to the 21st century* (pp. 50–63). Routledge.
Rodin, D. (2003). *War and self-defense.* Clarendon Press.
Rodin, D., & Shue, H. (2008). Introduction. In D. Rodin & H. Shue (Eds.), *Just and unjust warriors: The moral and legal status of soldiers* (pp. 1–18). Oxford University Press.
Schön, D. A. (1992). *The reflective practitioner: How professionals think in action.* Routledge.
Shaw, W. H. (2016). *Utilitarianism and the ethics of war*. Routledge.
Shue, H. (2008).Do We Need a Morality of War? In D. Rodin and H. Shue (eds.), Just and Unjust Warriors: The Moral and Legal Status of Soldiers (pp. 87–111). Oxford University Press.
Skerker, M. (2019). Preface. In M. Skerker, D. Whetham, & D. Carrick (Eds.), *Military virtues* (pp. xxv–xxvi). Howgate Publishing.
Skerker, M., Whetham, D., & Carrick, D. (Eds.). (2019). *Military virtues.* Howgate Publishing.
Steinhoff, U. (2012). Rights, liability, and the moral equality of combatants. *Journal of Ethics, 16*, 339–366.
Van Baarda, T. (2015). Between ethics and law: The ambiguous position of the state. *Journal of Military Ethics, 14*(2), 113–117.
Van Baarda, T., & Verweij, D. E. M. (Eds.). (2006). *Military ethics. The Dutch approach—A practical guide.* Martinus Nijhoff Publishers.
Van Doorn, G. (2019). Five critical success factors for coaching: A perspective on educating reflective practitioners. In W. Klinkert, M. Bollen, M. Jansen, H. de Jong, E. H. Kramer, & L. Vos (Eds.), *NL ARMS Netherlands (annual review of military studies 2019) educating officers: The thinking soldier—The NLDA and the Bologna declaration* (pp. 133–154). TMC Asser Press.
Van Wiggen, O. (2016). *Niemand is Belangrijker dan het Team.* White Elephant Publishing.
Verweij, D. E. M. (2020). *Ethiek en de Militaire Praktijk.* Boom.
Verweij, D. E. M., & Molendijk, T. (2019) Voorkomen is beter dan genezen: Moral injury vanuit macro- en micro-perspectief. In M. van der Giessen, P. H. Kamphuis, E. R. Muller, U. Rosenthal, G. Valk, & E. Vermetten (Eds.) *Veteranen: Veteranen en veteranenbeleid in Nederland* (pp. 165–184). Wolters Kluwer.
Waldron, J. (2018). Deep morality and the laws of war. In H. Frowe & S. Lazar (Eds.), *The Oxford handbook of ethics of war* (pp. 80–95). Oxford University Press.
Walzer, M. (2012). The aftermath of war: Reflections on *Just Post Bellum.* In E. Patterson (Ed.), *Ethics beyond war's end* (pp. 35–46). Georgetown University Press.
Walzer, M. (2015). *Just and unjust wars: A moral argument with historical illustrations* (5th ed.). Basic Books.

Prof. Lonneke Peperkamp holds the chair in Military Ethics and Leadership at the Netherlands Defence Academy, as the successor of Prof. Dr. Desiree Verweij. She is affiliated researcher at iHub (Radboud University Nijmegen) and King's College Centre for Military Ethics. Lonneke has an interdisciplinary background in law, philosophy, and conflict studies, and holds a PhD in just war theory (specifically *jus post bellum* and peace building). Her main research interests are war and political violence, military ethics, peace building, space law and security, global justice, and human rights. She is a member of the Board of Directors of EuroISME (the International Society for Military Ethics in Europe).

Dr. Christian Nikolaus Braun is a Radboud Excellence Initiative Fellow at Radboud University's Research Centre for State and Law (SteR). Previously, Christian was a Senior Lecturer in the Defence and International Affairs Department at the Royal Military Academy Sandhurst. He received his PhD from Durham University, UK. His primary area of research is the ethics of war and peace. Christian's research has been published in several peer-reviewed journals, including *Global Studies Quarterly, International Relations, International Theory*, the Journal of Military Ethics and the Journal of International Political Theory. His first research monograph, titled *Limited Force and the Fight for the Just War Tradition*, is forthcoming with Georgetown University Press.

Educating for Restraint

Peter Olsthoorn

Clausewitz made the intuitively appealing claim that wars tend to 'absoluteness', and that the limitations law and morality impose are in theory alien to it. Clausewitz of course knew that there are in practice many limitations to how wars are fought, but saw these restrictions as alien to what war is. Since then, historians such as John Lynn (2003), John Keegan (1993) and Victor Davis Hanson (1989) have taught us to see things differently: culture is central to understanding how wars are fought. Rituals and taboos set limits to what soldiers can and cannot do, and these limitations in fact form an essential element of what war is. A familiar example is the taboo on shooting at a lone soldier who forms too easy a target. This is the 'naked soldier' from Robert Graves' war memoirs, brought to fame by Michael Walzer's *Just and Unjust Wars* (1992; see also Chiu, 2019). Although such boundaries are as old as war itself, today it looks as if the limits to the violence militaries can use are stricter and more widespread than ever before. At present, these limitations spring more from the political and societal level than from the cultural and individual level. The law, politics, an increased moral sensitivity, extensive media coverage and public opinion, both at home and abroad, impose considerable (but mostly justified) limits on what troops can do.

As has been noted in many introductory paragraphs, the primary tasks of many militaries have shifted from national defence to the handling of international crises, ranging from humanitarian missions to outright war. These new operations, often at least partly undertaken for moral reasons, require a great deal of self-control on the part of military personnel. Having to function under the watchful eye of politicians, the media and the general public, ethics education for military personnel today partly comes down to convincing military personnel of the importance of exercising restraint, even when their opponents do not. Incidents in Afghanistan and Iraq have shown that the required moderation does not always come naturally. The

P. Olsthoorn (✉)
Netherlands Defence Academy, Breda, The Netherlands
e-mail: PHJ.Olsthoorn.01@mindef.nl

E.-H. Kramer and T. Molendijk (eds.), *Violence in Extreme Conditions*,
https://doi.org/10.1007/978-3-031-16119-3_9

killing of 39 civilians by Australian special forces in Afghanistan between 2005 and 2016 is a fairly recent example of such an incident (Inspector-General of the ADF, 2020), whilst the Abu Ghraib prison scandal and the Haditha massacre date back a little further into the past. In all of these three cases, the victims were outsiders to the military organization. Such innocent local civilians are the most visible victims of military misconduct. Of course, militaries also have to deal with serious misbehaviour *amongst* military personnel, but this is not very different from what we find in other organizations (although it is perhaps more prevalent in the military due to a number of specific characteristics). It is first and foremost the fact that the military can legitimately use violence that separates the military profession from almost all other professions. It is also what makes the ethical challenges for military personnel all the more testing, and underlines the importance of finding ways to prevent military personnel from crossing the line between legitimate force and unlawful violence.

Traditionally, militaries stressed the importance of obedience to rules and codes of conduct to that end, and clearly, pointing out what is permitted and what is not should have a role in any ethics education. Rules make clear to military personnel what actions are off limits. An example is the prohibition of torture, a ban that must be maintained regardless of how convenient it might be not to do so, and any flexibility here could bring us onto a slippery slope rather quickly. Similarly, we do not leave decisions concerning the use of certain types of weapons, such as chemical and biological ones, to the discretion of the individual soldier. Rule-based ethics point to the importance of having universal, categorically binding moral norms. This is not only in the interest of outsiders to the military organization but also in the interest of soldiers themselves. Research shows that military personnel who lack such rules experience more moral dilemmas, increasing the likelihood of moral injury (Schut, 2015). The drawback of rules is that they are often mostly ineffectual when there are no observers around. Moreover, rule-following can impede the ability to see the moral aspect of what one is doing, whilst that ability is evidently essential to morally sound decision-making. Perhaps the most important downside of such rule-based approaches, however, is that rules lack flexibility, also when that flexibility is clearly called for. Rules should therefore leave soldiers with some leeway in decision-making, if only to keep them from committing what have been referred to as 'crimes of obedience' (Kelman & Hamilton, 1989). It is perhaps for that reason that one textbook on military ethics explicitly states that 'in any situation where law and ethics set different standards, a member of the military profession will follow the higher standard, inevitably the one required by ethics' (Coleman, 2013, p. 268).

Making good use of this leeway presupposes a good disposition, and many militaries for that reason see a virtue-based approach to teaching military ethics as a necessary complement to rules imposed from above in their effort to ensure that military personnel exercise restraint in their use of force. Virtue ethics is in keeping with the tendency of many militaries to move away in their ethics education from a largely functional approach that is mainly about military effectiveness towards a more aspirational approach that focuses on character and aims at making soldiers better persons. This shift is mainly based on the view that bad persons are not likely to form morally good soldiers, although they could of course still be effective ones:military history is replete with examples of military leaders who were effective but absolutely not ethical (Robinson, 2007a; Wolfendale, 2008, p. 164). What makes virtue ethics especially interesting for the military is its premise that character can be developed, and that virtues are not to be understood as innate qualities but as dispositions that can be acquired through training and practice. Such an approach sits rather well with the way most militaries see themselves: as being in the business of character-building. Many militaries have adopted this aspirational virtue ethics approach in a rather carefree manner, however, more or less overlooking the complexities that come with this approach.

In reality, there are quite a few unanswered questions. To state a few: how do we teach virtues? It is an assumption of virtue ethics that they can be taught, but is this really the case? And if so, how should they be taught? And at what age? If character is formed before one enters the military, this presents a problem for a military ethics curriculum founded on virtue ethics. What is more, virtues are allegedly developed by practising them, but all too often military ethics education consists of formal education in a classroom setting that leaves little room for that. Does ethics education based on virtue ethics not often consist of teaching *about virtues* (and virtue ethics) rather than teaching virtues? The most important question, however, is *which* virtues should form the building blocks for a virtue-based education.

The answer to the question as to which virtues military personnel need today depends at least in part on the answer to the underlying question as to whether military virtue is independent of place and time. At first sight, a convincing argument can be made that this is indeed the case, at least to some extent. Some military virtues are valued in all eras and cultures, for instance because they perform an important function in or for the military. Physical courage is, of course, the obvious example here, being the archetypal military virtue. But if we take a closer look at courage, it also becomes clear that the type of courage that is needed, and even what we mean by the term courage itself, is subject to change over time. Physical courage has always been less of a virtue for military personnel in units that are in fact never deployed, the so-called 'cold organization' (Soeters et al., 2003). Furthermore, the rise of a number of new technologies make physical courage also obsolete for at least some 'hot' parts of the military. Cyber soldiers and UAV operators, for instance, do not seem to need this type of courage at all.[1] Some years ago, Jesse Kirkpatrick (2015a, 2015b) and Robert Sparrow (2015) had a thought-provoking although somewhat semantic discussion on the courage of drone operators. If a conclusion had to be drawn from that discussion, it would be that these operators do need courage, but less in the form of martial courage than of moral courage. We return to this point later. Regarding other traditional, central military virtues, such as loyalty, discipline and obedience, it is at a minimum less clear what positive role they could have for, say, cyber operations or operating armed drones. More worryingly, the virtues

[1] An article published in *The New York Times* (Schmidt, 2016) a few years ago describes how, for a long time, drone operators were viewed 'more as video game players than as warriors. But in a reflection of their increasingly important role under President Obama, the drone operators will now be eligible for military honors akin to those given to pilots who flew over the battlefields of Iraq and Afghanistan'.

that prevail in most militaries are mainly beneficial to the interests and aims of the organization and colleagues (Olsthoorn, 2010). With the possible exception of respect, which at least some militaries list as a virtue, these virtues are not particularly helpful to the local population of the countries that military personnel are deployed to. Instrumental in attaining the objectives of the military, there is little in them that limits the behaviour of soldiers towards civilians.

Assuming that the traditional martial virtues such as physical courage do not always suffice today, there are at least three possible answers to the question as to what we do need instead. One could argue, first of all, that the virtue approach is still the best one but that we need virtues that are just better suited for today than the traditional, rather bellicose ones are. But one could also argue, secondly, that not only is the virtue approach the right one, the traditional virtues are by and large the virtues that we still need—with the caveat that the new operations require new interpretations of these virtues. A third option is that we start looking for something different altogether, most probably a more rule-based or utilitarian approach to ethics education, or a combination of both.

Looking for 'New' Virtues

If we assume that the virtues militaries traditionally try to espouse are of limited use in regulating the conduct of military personnel in today's conflicts, devising a new list of virtues would be a first possible way ahead. In this line of thought, today's soldiers do indeed need virtues, but not necessarily of the 'duty, honour, country' variety that prevails at present. As said, the virtues we teach military personnel should fit their responsibilities, and the virtues needed today are most likely more about exercising restraint than about demonstrating physical courage, loyalty and discipline. Virtues of restraint, although very relevant for military personnel, would be less military-specific in the sense that they would be closer to the virtues valued by society at large. Incorporating such virtues of restraint could therefore bring the military into closer alignment with wider society. Opting for a set of virtues that is closer to what we could call 'common morality' would also fit the more aspirational and less functional approach that militaries are moving towards in their ethics education.[2] The 'general' cardinal virtues of course form a natural source to turn to first when looking for aspirational, comprehensive virtues. Interestingly, of the four cardinal virtues of courage, wisdom, temperance and justice, only courage has until now made it to the traditional lists of military virtues. The equally cardinal virtues of

[2] There is sometimes a difference between mainly functional role morality and more aspirational common morality: we expect lawyers to defend the guilty, and spies may use deceit (Coleman, 2013). Although role morality clearly differs from ordinary morality also for military personnel, we have already noted that there is a tendency in many Western militaries towards a less functional approach. One could also argue, however, that by aiming to instill both 'general' virtues, such as integrity and honesty, and more military-specific virtues, such as courage and discipline, the military combines an aspirational and a functional approach.

wisdom, temperance and justice, today probably at least as needed as courage is, are absent on most lists of military virtues (although they do surface in a recent book on military virtues; see Skerker et al., 2019). That is to be regretted, as wisdom, justice and temperance are clearly more encompassing than the traditional military virtues are, and are a lot more relevant when it comes to exercising restraint. Opting for the cardinal virtues would also give us a set of virtues that does justice to the now nearly forgotten ancient meaning of the word integrity, according to which all of the virtues are interrelated and one therefore cannot possess one virtue without having the others too. Being just is of little value if one lacks the courage to defend what is just, for instance. Likewise, courage is of not much use without practical wisdom to guide it, whilst that same courage is not a virtue if it does not serve a just goal. Wisdom uninformed by justice may come close to cunning. The rather jumbled collections of virtues that militaries now advocate (see also Robinson, 2008) lack these important interconnections, and at times the listed virtues even appear to contradict each other. Loyalty and integrity, for instance, are two virtues that will conflict on occasion. Military whistle-blowers, for example, choose integrity over loyalty, but often pay a heavy price for that because their (former) colleagues and the organization deem them disloyal.

To complicate matters, what, on the face of it, pretty straightforward virtues such as justice and wisdom stand for is rather time and place dependent. For instance, if we take a closer look at justice, we see that the classical understandings of that virtue ('to each his own') were much more inegalitarian than our current interpretation of what is just allows for.The underlying hierarchical worldview and corresponding ideas about justice motivated Aristotle's infamous defence of slavery, for instance. Apparently, our arguments against slavery are mainly convincing to those who subscribe to the modern idea that all people are equal, a notion that was alien to Aristotle and his contemporaries.Whilst Aristotle believed that he had given an objective description of moral and intellectual virtues that were rooted in a shared human nature, he had in fact mainly described the qualities that an Athenian gentleman of the fourth century BC would ideally possess.

That the way we interpret virtues makes such a difference is not only a complication, it also presents us with a second way forward. One could reason that it perhaps suffices to identify the weaknesses of the existing military virtues and find ways to interpret these traditional virtues in a manner that does not suffer from these pitfalls, and that formulating a new list of more outward looking and less bellicose virtues is hence unnecessary (and perhaps a bridge too far for the relatively traditional organization that the military is). The question is then not which new virtues the military should promote, but in what form the existing ones should best be understood.

Interpreting Old Virtues in New Ways

Although militaries today mostly cling to traditional interpretations, other readings of the military virtues are of course possible. A second way forward would therefore be to interpret the existing virtues somewhat differently; that is, less narrowly than is commonly the case (see also Schulzke, 2016, pp. 195–196). The virtue of courage is especially interesting in this context. Most definitions of courage in the military still hark back to Aristotle, who defined courage as the mean between rashness and cowardice (*Nicomachean Ethics* 1115). This idea of courage fitted the hoplite warfare of his day very well, as both an excess or a deficiency of daring would destroy the organized whole that the phalanx was. But this martial notion of courage on the battlefield could clearly not be further away from what Gandhi imagined when he pleaded for courageous but peaceful resistance to the British colonial power. Clearly, the term courage can denote different things in different settings, and where Aristotle wrote about physical courage, Gandhi called for a type of courage that we commonly call moral courage.

Moral courage is an important subspecies of the virtue of courage, as it asks us to stick to our principles even if others disagree and perhaps hold us in contempt for upholding them. Physical courage is primarily something one's superiors and colleagues benefit from. Moral courage has a wider reach, and is, in line with the aforementioned distinction between an aspirational and a functional approach, more about being a better person than about being an effective soldier (compare Robinson, 2007a, p. 22; Robinson, 2008, p. 1). Today, its beneficiaries are not only military colleagues, as is predominantly the case with physical courage, but also the outsiders that the military is there to protect. It was suggested above that drone operators particularly need moral courage, and Peter de Lee (2019) gives an excellent example of just that when he describes how an acting sergeant on her first day in a supervisory role overseeing a Reaper team stuck to her judgement, against the opinion of all present, that an alleged parcel placed on the back seat of a motorcycle being used by a Taliban target was in fact a child, which in the end it did indeed turn out to be.[3] It is such wider interpretations of the military virtues that can provide guidance to military personnel in morally ambiguous situations, as providing general rules will not work in complex situations.

On a positive note, many contemporary interpretations of military courage already include moral courage. However, it can only perform its important role if militaries allow room for it. Yet although militaries today rarely fail to at least pay lip service to moral courage, in reality military organizations can on occasion offer a fairly unfriendly environment for the morally courageous, especially when adherence to principles conflicts with organizational interests or mission success. We already briefly noted above that loyalty, as most militaries define it, conflicts with integrity, which is also regularly listed as a military virtue. Many a whistle-blower can testify to that: they tend to end up as martyrs for the good cause, not as moral examples for their military organization. The explanation for this sobering fact lies in the interpretation that most militaries give to the virtue of loyalty.

---

[3] She would have been equally courageous, of course, if the supposed parcel had turned out to be precisely that: a parcel. In that case, however, it would perhaps have been more difficult to muster that same amount of moral courage at another time.

Loyalty is included on lists of military virtues as frequently as courage is. The type of loyalty that most military organizations foster, however, is mainly confined to loyalty to the organization and colleagues. The military is not alone in this interpretation of loyalty as group loyalty. In general, loyalty tends to signify in some way giving priority to the interests of an individual, a group or a country, even when reason dictates a different direction (Ewin, 1992, p. 406). This is the form of group loyalty that militaries tend to promote in education and, especially, socialization. Disloyalty is, from that point of view, a very serious offence that adds insult to injury. However, someone taking a less partial standpoint might argue that the aforementioned whistle-blowers are not only disloyal but may even qualify as loyal, albeit to a principle instead of to a group or an organization. Different from group loyalty, loyalty to principle does not require the suspension of independent judgement. From this point of view, loyalty to principle might qualify as a virtue, whereas group loyalty, with its partiality to the near and dear, probably would not.

In the military, interpreting loyalty as group loyalty causes it to function as a root of both unethical conduct and attempts to cover that conduct up.[4] Military training reinforces this, being sometimes aimed more at furthering group loyalty than at cultivating autonomous individuals. As a result, military personnel usually identify mainly with the small group of colleagues with whom they spend most of their time. It would be a significant improvement if militaries would understand loyalty to include loyalty to a profession or principle, not just to a group and an organization (see also Olsthoorn, 2011). Loyalty to one's professional ethic, instead of to one's organization and colleagues, is what is commonly understood to be one of the key characteristics of a professional, something most military personnel claim to be.[5] Especially at a time when many armed forces consider the promotion of universal principles as their main ground for existence, and on occasion even claim to be 'a force for good', the development of a truer professionalism, with the main focus of

---

[4] A report on the Netherlands Defence Academy observes that 'the great value that is attached to loyalty, group formation and comradeship and the intensive training that military personnel undergo together' can 'sow the seeds for a military practice in which there is an excessive inward focus' (2014, p. 10). The main reason cadets give for their unwillingness to report incidents is 'the idea that it is not in keeping with comradeship, that it is disloyal' (COID, 2014, p. 18). Interestingly, group loyalty here requires something different (i.e. not reporting) than loyalty to the organization. Somewhat ironically, most cadets said that later, when they were in leading positions, they would want their subordinates to report misbehaving colleagues—organizational loyalty should then trump loyalty to colleagues. A recent report on social safety within the Netherlands Defence organization as a whole similarly found that loyalty to the group reduces the willingness to report incidents (Giebels et al., 2018, p. 65). According to the report, the organizational culture with its emphasis on loyalty is an important cause of a lack of social safety (2018, p. 7).

[5] Loyalty to the organization is the main aspect of military professionalism that is somewhat at odds with what a 'regular' professional ethic entails. Armed forces thoroughly socialize their employees into the organization, which contributes to the strong loyalty military personnel feel towards each other and their employer. That military personnel are predominantly trained in house makes this socialization into the organization easier. As a consequence, different militaries have different organizational values (often still service specific), but there are no values of the military profession as such. By contrast, the values and standards of regular professionals stem from universities and professional associations, not from, for example, their hospital or law firm.

loyalty being the soldier's professional ethic instead of his or her organization, would be a step forward.

Finally, let us take a look at a virtue that is at first sight not a very martial one, namely that of respect. Respect appears on the value lists of several militaries, and was mentioned above as a virtue that both the military and society at large value. A closer look, however, reveals that respect in the military is at times limited to respect towards colleagues. The US Army, for instance, describes respect as, amongst other things, 'trusting that all people have done their jobs and fulfilled their duty', adding that '[t]he Army is one team and each of us has something to contribute'.[6] This definition seems to tacitly limit respect to colleagues. Military ethicist Timothy Challans describes how 'early drafts of the Army's 1999 leadership manual included the notion of respect; in fact, the key feature of respect was that of respecting the enemy on the battlefield. That idea did not survive the staffing process, and even a cursory check of the manual today will reveal that only Americans are mentioned as being recipients of this important value of respect' (2007, p. 163).[7]

Listing respect as a virtue is therefore not the concession to the current tasks of the military that it might seem: although 'respect' certainly sounds inclusive, at present, the way some militaries interpret the term bars it from being that. Such interpretations fail to take into account that military personnel will often be doing their jobs amidst the local population. As one author stated, somewhat boldly, 'non-soldiers lie outside the military honour group; as such they are felt to deserve no respect' (Robinson, 2007b). Why exactly militaries are reluctant to include outsiders remains somewhat of a mystery, given that respect is not a constant-sum game; respect for outsiders does not reduce the amount of respect left to show colleagues. Even if it is true that colleagues, not outsiders, are those who in fact suffer most often from misconduct in the military, this exclusive attention for their well-being seems a bit too one-sided.[8] Here, too, more inclusive interpretations seem justified.

---

[6] The US Army values can be found at https://www.army.mil/values/.

[7] Somewhat similarly, the Dutch military published a new code of conduct in 2006 that contained the sentence 'I treat everyone with respect'. A look at the accompanying explanation showed that the pronoun 'everyone' referred exclusively to colleagues who should be safeguarded against harassment, sexual intimidation and discrimination. In 2018, that code of conduct was replaced by a new one. The new code also only regulates the behaviour of military personnel towards each other, not their behaviour towards outsiders.

[8] However, there is another side to this: Western military personnel sometimes face situations 'in which the conduct of the local population in a deployment area (a different culture) [was] experienced as conflicting with one's own personal moral and cultural values' (Schut, 2015, p. 106). Pre-deployment training teaches Western military personnel respect for other cultures, as Western forces can be involuntarily offensive in their dealings with the local population if they have 'a lack of cultural relativity in their occupation "technique"' (Fontan, 2006, p. 219). However, emphasizing the need to respect other people's mores provides Western soldiers with a reason for not intervening in cases of corruption or the abuse of women and children. A soldier deployed to Afghanistan explained: 'During Mission- specific Training, we didn't discuss this subject at all. But we did learn that we must respect local culture' (Schut, 2015, p. 116).

Alternatives for a Virtue Approach

A final way ahead would be to reconsider whether virtues form the best underpinning for the ethics education of military personnel in the first place. We already saw that the traditional military virtues are in themselves more inward looking than the cardinal virtues. However, also on more a theoretical level, virtue ethics is fairly self-regarding: virtue ethics focuses on the agent and his or her character and flourishing, even in situations (and war is probably such a situation) where an outcome-centred approach would seem to be more appropriate. The aim of virtue ethics is one's own flourishing. It is therefore less attentive to the needs of others than is, for instance, the utilitarian notion of the greatest happiness for the greatest number, or the rule-based maxim that one should treat others as one would like others to treat oneself. Interestingly, the fact that militaries promote virtues with an eye to external goals such as military effectiveness or ensuring the ethical use of force raises the question whether it is virtue ethics that is being practised here in the first place. Promoting certain virtues because they are beneficial to others within or outside one's own organization amounts to what is sometimes described as character utilitarianism (Railton, 1988).[9]

That brings us to utilitarianism: next to rules and virtues it is in theory a possible third candidate to buttress military ethics education, as it is a universalistic ethic that holds that everyone's life and happiness should weigh equally. In practice, most authors on military ethics see it as particularly unfit for that purpose, mostly because it would make military expedience outweigh all other concerns: 'an outcome-centered approach may lead all too easily to military expedience as the sole guide to actions in war' (Bonadonna, 1994, p. 18). However, utilitarianism does not condone the maximizing of our own utility, as some critics seem to hold, but that of all. This means that the utilitarian dictum that the consequences to all persons should weigh equally would, if taken seriously, lead to a fairer distribution of the right to life (see also Shaw, 2016). From a utilitarian viewpoint, one could for instance argue that soldiers should take as much care to avoid casualties amongst enemy civilians as they do for their own civilians. Although such an impartial view may be expecting too much from regular soldiers in a regular war in defence of one's own country, in many of today's operations military personnel probably should be able to do so a bit more easily.

In general, the aspirational approach focuses on character, whilst the functional approach is based more on conduct and outcomes. This corresponds with three main schools in moral philosophy, namely virtue ethics, rule-based ethics and utilitarianism. An alternative to the usual virtue ethics, rule-based ethics and utilitarianism advance the idea of taking values as a basis for ethics education for military personnel. Virtues and values are two different things, even if militaries sometimes treat them as if they were the same. Virtues represent 'desirable characteristics of individuals, such as courage', whilst values, on the other hand, correspond to 'the ideals that the

[9] Some argue that it is morally dubious to mould someone's character with the aim of making him or her a better soldier (Robinson, 2007a, p. 32).

community cherishes, such as freedom' (Robinson, 2007a, p. 32). A value-based ethics education could put the values and principles of that community in the foreground instead of promoting the military-specific virtues that are currently taught. An example of such a value-based approach is the concept of *Innere Führung* as used by the German armed forces. A leading idea behind *Innere Führung* is that soldiers should think for themselves. The values that should guide that independent thinking, however, are societal: members of the German military are to actively defend, out of personal conviction, values such as human dignity, freedom, justice, equality and democracy. The aim is to bring the military into alignment with civil society. It might also bring the values of the military somewhat more into alignment with the humanitarian ideals underlying many of today's operations.

But where does that leave us? Does a value-based approach just add a fourth alternative to choose from? Most moral philosophers tend to have a clear preference for one of the approaches outlined above. In real life, however, most of us tend to see a role for virtues, values and rules, whilst also taking the consequences of an act into consideration. We are probably right to do so, as none of these schools has the ultimate answer. It seems that those involved in professional ethics education are practically duty bound to adopt a similar mixed approach. In philosophy, one finds 'parts of the truth (along with much error) everywhere, and the whole truth nowhere' (Appiah, 2006).

Discussion

Opting for such a mixed approach, however, does not alter the fact that it is uncertain whether ethics education for uniformed personnel has any tangible beneficial effects on their conduct. We know little about what works and what does not in military ethics education. It almost certainly augments the moral awareness of military personnel, but this does not necessarily mean that it also directly contributes to better behaviour. Perhaps the positive effects of military ethics education are indirect; that is, providing formal ethics education improves the ethical climate and in the long run therefore also the behaviour of military personnel. This is mere conjecture, however, and given the amount of time and effort spent on ethics education, the question of whether it works deserves more consideration. A good first step would be to think more systematically about the military virtues that we want to teach. As it stands, many publications on military virtues deal with one specific virtue only, such as courage or loyalty, whilst broader approaches that go into the relationships between the different virtues are relatively rare. These publications do not refer to much scholarly literature and are as a result sometimes rather uncritical, as they mainly stress the importance of a particular virtue and not so much its complexities. That is a pity, as it is clear that, as currently interpreted, some of the traditional martial virtues, such as courage and loyalty, are less relevant today. For that reason, we need to look for alternatives to the traditional military virtues. A few possible ways forward have been outlined above.

References

Appiah, K. A. (2006). *Cosmopolitanism: Ethics in a world of strangers*. W.W. Norton & Company.

Aristotle. (1962). *Nicomachean ethics*. The Bobbs-Merrill Company.

Bonadonna, R. R. (1994). Above and beyond: Marines and virtue ethics. *Marine Corps Gazette, 78*(1), 18–20.

Challans, T. L. (2007). *Awakening warrior: Revolution in the ethics of warfare*. State University of New York Press.

Chiu, Y. (2019). *Conspiring with the Enemy: The ethic of cooperation in warfare*. Columbia University Press.

COID. (2014). *Integriteit bij de opleiding en vorming van adelborsten en cadetten aan de Nederlandse Defensie Academie: Duiding van risico's, perspectief op verbetering*. Netherlands Ministry of Defence.

Coleman, S. (2013). *Military ethics*. Oxford University Press.

Ewin, R. E. (1992). Loyalty and virtues. *Philosophical Quarterly, 42*(169), 403–419.

Fontan, V. (2006). Polarization between occupier and occupied in Post-Saddam Iraq: Colonial humiliation and the formation of political violence. *Terrorism and Political Violence, 18*(2). https://doi.org/10.1080/09546550500383266.

Giebels, E., van Oostrum, F., & van den Bos, K. (2018). *Commissie Sociaal Veilige Werkomgeving Defensie, eindrapportage: Onderzoek naar een sociaal veilige werkomgeving bij Defensie*. Netherlands Ministry of Defence. https://www.defensie.nl/downloads/rapporten/2018/10/15/eindrapport-onderzoek-naar-een-sociaal-veilige-werkomgeving-bij-defensie.

Hanson, V. D. (1989). *The western way of war*. Oxford University Press.

Inspector-General of the Australian Defence Force. (2020). *Afghanistan Inquiry Report*. https://afghanistaninquiry.defence.gov.au/sites/default/files/2020-11/IGADF-Afghanistan-Inquiry-Public-Release-Version.pdf.

Kelman, H. C., & Hamilton, V. L. (1989). *Crimes of obedience: Toward a social psychology of authority and responsibility*. Yale University Press.

Kirkpatrick, J. (2015a). Drones and the martial virtue courage. *Journal of Military Ethics, 14*(3–4), 212–219. https://doi.org/10.1080/15027570.2015.1106744.

Kirkpatrick, J. (2015b). Reply to sparrow: Martial courage—Or merely courage? *Journal of Military Ethics, 14*(3–4), 228–231. https://doi.org/10.1080/15027570.2015.1109942.

Lee, P. (2019). Case Study 2: Moral courage. In M. Skerker, D. Whetham, & D. Carrick (eds.), *Military virtues*. Howgate Publishing.

Lynn, J. (2003). *Battle: A history of combat and culture*. Westview Press.

Keegan, J. (1993). *A history of warfare*. Random House.

Olsthoorn, P. (2010). *Military ethics and virtues: An interdisciplinary approach for the 21st century*. Routledge.

Olsthoorn, P. (2011). Loyalty and professionalization in the military. In J. Wolfendale & P. Tripodi (Eds.), *New wars and new soldiers: Military ethics in the contemporary world* (pp. 257–272). Ashgate.

Railton, P. (1988). How thinking about character and utilitarianism might lead to rethinking the character of utilitarianism. *Midwest Studies in Philosophy, 13*, 398–416. https://doi.org/10.1111/j.1475-4975.1988.tb00135.x.

Robinson, P. (2007a). Ethics training and development in the military. *Parameters*, Spring, 22–36.

Robinson, P. (2007b). The way of the warrior. *Spectator*, 13 June.

Robinson, P. (2008). Introduction: Ethics education in the military. In P. Robinson, N. de Lee, & D. Carrick (Eds.), *Ethics education in the military* (pp. 1–12). Ashgate.

Schmidt, M. S. (2016). Pentagon will extend military honors to drone operators far from battles. *The New York Times*, 6 January.

Schulzke, M. (2016). Rethinking military virtue ethics in an age of unmanned weapons. *Journal of Military Ethics, 15*(3), 187–204. https://doi.org/10.1080/15027570.2016.1257851

Schut, M. (2015). *Soldiers as strangers: Morally and culturally critical situations during military missions* (Doctoral dissertation), Radboud University.
Shaw, W. H. (2016). *Utilitarianism and the ethics of war*. Routledge.
Skerker, M., Whetham, D., & Carrick, D. (Eds.). (2019). *Military virtues*. Howgate Publishing.
Soeters, J., Winslow, D., & Weibull, A. (2003). Military culture. In G. Caforio (Ed.), *Handbook of the sociology of the military* (pp. 237–254). Kluwer Academic/Plenum.
Sparrow, R. (2015). Martial and moral courage in teleoperated warfare: A commentary on Kirkpatrick. *Journal of Military Ethics, 14*(3–4), 220–227. https://doi.org/10.1080/15027570.2015.1109946
Walzer, M. (1992). *Just and unjust wars*. Basic Books.
Wolfendale, J. (2008). What is the point of teaching ethics in the military? In P. Robinson, N. de Lee, & D. Carrick (Eds.), *Ethics education in the military* (pp. 161–174). Ashgate.

Dr. Peter Olsthoorn is Associate Professor in Military Leadership and Ethics at the Netherlands Defence Academy. Besides leadership and ethics, he teaches on armed forces and society, war and the media, and on ethics and fundamental rights in the European Joint Master's programme in Strategic Border Management. His research is mainly on topics such as military virtues, military medical ethics, armed drones, and the ethics of border guarding. Among his publications are the books *Military Ethics and Virtues: An Interdisciplinary Approach for the 21st Century* (Routledge, 2010) and *Honor in Political and Moral Philosophy* (State University of New York Press, 2015).

The E-Word (Emotions) in Military Ethics Education: Making Use of the Dual-Process Model of Moral Psychology

Edgar Karssing

According to Verweij (2016), military ethics education should be more than learning how to reason by using ethical theories.[1] Explicit attention should be paid to the 'e-word': emotions. Verweij uses the diminutive 'e-word' to indicate that considering emotions is uncomfortable for many people: "emotions determine the way we think and act, yet, at the same time there is a mistrust of emotions or at best an ambivalent attitude towards these so called 'irrational aspects', or 'blind forces' as they are often perceived, not only in a military setting, but by many philosophers as well" (Verweij, 2016, p. 28). Consequently, giving emotions a proper role in ethics education is challenging. Yet, it is a challenge that should be taken as "a good functioning of emotions and feelings is necessary for social and humane behavior" (Verweij, 2016, pp. 41–42). For it is through emotions that humans are able to experience values as meaningful: "when people have no emotions, when there is no sensitivity, or when people are dispassionate or numb, values will have no meaning to them and are perceived as unfamiliar words on paper that other people seem to fuss about. Obviously, this may have far-reaching and undesired consequences in military practice, for soldiers in these circumstances easily cross moral barriers. This is not only to the detriment of the people these soldiers are confronted with but often also affects the soldier himself, as studies on PTSD and 'moral injury' indicate" (Verweij, 2016, p. 28).

According to Verweij, the model of Haidt makes an important case for emphasizing the 'e-word' with regard to ethics education. Haidt is an important pioneer within moral psychology of the dual-process model for understanding moral judgment. In this model, moral emotions play an important role. Yet, throughout the model, Haidt attributes only a minuscule role to the act of reasoning. This clashes with Verweij's views on education, as she sees practice with reasoning aided by the

E. Karssing (✉)
Nyenrode Business Universiteit, Breukelen, The Netherlands
e-mail: e.karssing@nyenrode.nl

[1] This chapter is an elaboration of a preliminary study, see Karssing (2021).

E.-H. Kramer and T. Molendijk (eds.), *Violence in Extreme Conditions*,
https://doi.org/10.1007/978-3-031-16119-3_10

help of philosophical insights as a precondition for meaningful education. Otherwise, a conversation about moral issues does not get much further than "an exchange of opinions never leading to the vital reflection, necessary for adequate moral decision-making" (Verweij, 2016, p. 27).

This chapter will first discuss the dual-process model in a general sense. Then, the consequences linked to it by Haidt will be discussed. The dual-process model has been embraced by many and indeed provides tools for embracing the 'e-word' in ethics education in practical ways. However, it is first necessary to take some distance from specific elements of Haidt's views. This will be done by introducing the perspective of Musschenga, who does not reject Haidt's model, yet sees it as too one-sided. This will be followed by a discussion of Musschenga's criticism of the model. Finally, different methods to give the e-word a firm place in ethics education will be discussed.

Haidt and the Dual-Process Model in Moral Psychology

Anyone concerned with ethics education, with strengthening the moral competence of professionals, knows that you cannot limit yourself to philosophy. Rather, you should also look closely at the behavioral sciences that study how people develop morally and how they (can) deal with moral issues. The behavioral sciences provide the empirical knowledge that is necessary to determine what works: the theory must fit the practice (cf. Flanagan, 2017).

During the last decades of the previous century, the theory of psychologist Kohlberg was dominant. According to his theory, people develop morally by passing through several stages. Each of these stages is characterized by a specific way of thinking about moral issues, and in the final stage, morally mature people are able to make moral judgments entirely independently. The description of this last stage fits well with philosophers such as Kant and Rawls, who emphasized reasoning, but had little use for emotions. Using Kohlberg's theory, it is possible to explain the importance, and effectiveness, of discussing practical examples with professionals in order to strengthen their thinking and reasoning skills. However, over the last twenty years, a small revolution has taken place. A 2001 article by psychologist Haidt is often pointed to as the start of this revolution as, throughout the article, Haidt fundamentally undermines Kohlberg's theories (Haidt, 2001, cf. Ellemers et al., 2019). Haidt emphasizes intuition and emotions as informing moral judgments, with reasoning only playing a small role. In particular, reason behaves like a lawyer who devises arguments after the fact to justify positions that have been taken based on intuition and emotions.

Haidt's theory fits within a broader research program in moral psychology focused on the nature of moral judgments. The words 'research program' are a deliberate choice: although there is a clear affinity between Haidt and his colleagues, they do not necessarily all subscribe to the same theory, neither do they conduct research in the same way, nor do they reach the same conclusions (cf. Brand, 2016). However, Haidt and his colleagues do all make use of the dual-process model of moral judgment. The model itself has been adopted from the behavioral sciences and has become best known among the general public through the book *Thinking, fast and* slow by Nobel Prize winner Kahneman (2011). The basic model will now briefly be introduced before it is applied to moral judgment.

According to the dual-process model, a distinction can be made between two ways of making judgments: judgments based on intuition, and judgments based on deliberate thinking. Or, as coined by psychologists Stanovich and West and popularized by Kahneman, judgments based on System 1 and judgments based on System 2. It should be stressed that this is a way to paint a picture of the inner workings of our brain: these are metaphors, there are not actually two systems identifiable in our brain.

System 1 works intuitively. That is, implicitly, unconsciously and emotionally. It acts automatically, quickly, and with little or no effort. Consequently, it knows no doubt.

System 2 works through deliberate thinking. It works slowly, consciously, laboriously, explicitly, and logically. It consists of thoughtful deliberation and involves doubt.

Most of the choices made by humans are determined by System 1. System 2 is usually in a dormant state: people would rather be lazy than tired.

Thus, the model indicates that there are two different types of judging, intuitive judging and deliberate judging, that people apply. In the research program on moral judgments, this model is used as a framework from which to build other ideas. When it comes to System 1, moral intuitions, judgments that immediately impose themselves on us without further thought, are central. These intuitions indicate how a situation should be interpreted. They serve to answer questions such as: what is going on in this situation and what is my role in it, what is proper and what is improper, and what action is appropriate in this situation? In short, intuitions simultaneously point to a problem and a solution without the need for thought. Therefore, System 1 serves as a moral compass that leads the way when facing a moral question. Here, emotions play an important role. Intuitions are both cognitively and affectively charged, with feelings serving as a rule of thumb for arriving at quick moral judgments. Consequently, something is morally appropriate if, and because, it feels good. Similarly, something is inappropriate if, and because, it feels bad (Sinnott-Armstrong et al., 2010; Slovic et al., 2007). Conversely, the thinking and reasoning skills that are usually the focus during ethics education belong to System 2. Yet, in everyday life, judgments are more often made intuitively rather than well-considered.

Within the research program there is still a lot of discussion about definitions, about the interpretation of results, and about the way the research is conducted. Research often focuses on extreme examples that have little to do with everyday situations (cf. Brand, 2016; Sauer, 2019). Also, it has already been suggested that there may be a third system (Sauer, 2019). Additionally, there are different views on the relationship between the two systems: is there subordination (one system is more decisive in judging than the other), competition, or cooperation (cf. Brand, 2016; Liao, 2011; Sauer, 2019)?

Haidt is an important representative of the dual-process model and clearly outlines the role that intuitions, emotions, and moral reasoning play in moral judgments. His view is briefly summarized in the following prompts (cf. Brand, 2016; Greene & Haidt, 2002; Haidt & Bjorklund, 2008; Haidt & Kesebir, 2010; Haidt, 2001, 2012; Musschenga, 2008, 2009, 2010, 2011; Sauer, 2019).

- Our intuitions and emotions (System 1) are partly innate and partly learned (Haidt provides different evolutionary explanations for humanity's innate intuitions and emotions).
- System 1 usually prevails over System 2. Following the philosopher Hume, reason is the slave of emotions (and therefore intuitions). Haidt qualifies this statement by emphasizing that this is usually the case: "intuitive primacy (but not dictatorship)" (Haidt & Kesebir, 2010).
- In most cases, System 2 is set up as an advocate rather than the cool and critical thinker presupposed by ethical theories: System 2, as an advocate, provides post-hoc (i.e., after-the-fact) arguments for the judgment already determined by System 1. Therefore, reasoning hardly plays a role in making judgments. Instead, it is focused on justifying judgments that have already been made.
- System 2 can be used to think critically ('no dictatorship'), but this hardly ever happens. When it does happen, it is usually in complex situations when intuitions conflict. Critical thinking hardly motivates either: the motivation to act on judgments is informed by emotions, ergo System 1.
- For most people, System 2 will only play a critical role in a social setting. People are hardly capable of making critical and thoughtful judgments on their own; for that they need others to point out blind spots and new perspectives. Yet, also in a social setting, arguments will seldom affect moral judgments through the logic of system 2. Instead, arguments indirectly influence judgments by appealing to latent intuitions in System 1. In other words, because the other person knows how to press the right emotional buttons of System 1.
- Critical thinking in a social setting is 'biased'. People tend to conform to the views of their friends and are much more critical of arguments that are at odds with their intuitive judgments than of arguments that actually confirm what they already believe.

In conclusion, according to Haidt, reasoning hardly plays any role in moral judgments. Consequently, there is in ethics education little point to professionals discussing practical examples with each other, using insights from philosophy. According to Haidt, people mainly judge intuitively and emotionally: "sometimes these affective reactions are so strong and differentiated that they can be called moral emotions" (Haidt & Kesebir, 2010). These moral emotions can be cultivated to a limited extent, but this is mainly done in social settings.

Haidt Revisited

According to Verweij (2016), emotions need more emphasis in military ethics education, with the precondition that philosophical insights also receive attention. Haidt shows that emotions do indeed play an important role in moral judgments through System 1. However, he hardly leaves any room for reasoning. Let alone for the use of philosophical insights to reinforce moral judgments. Musschenga (2009, 2010, 2011) endorses Haidt's conclusion that people very often make intuitive judgments based on their feelings. Additionally, he advocates that this should be an important starting point in ethics education. However, Musschenga also has several criticisms on Haidt's work and certainly sees room for reasoning and philosophical insights. Briefly summarized, these criticisms amount to Haidt overestimating the role of System 1 and underestimating to role of System 2. According to Musschenga, because there is reason to be doubtful about the reliability of System 1, System 2 is needed to monitor, test, and correct System 1 where necessary. Essentially, although System 2 only rarely plays a leading role, the fact that System 1 can be unreliable means that System 2 must play a bigger and more important role than Haidt suggests.

An Increased Role for System 2

Why does Haidt understate the role of moral reasoning, the activities of system 2? Firstly, it is true that people regularly act as advocates for their own views in a discussion. This is especially true when it comes to topics on which they have already taken a firm position. Yet, many conversations concerning moral issues are precisely about topics on which one does not yet have a clear opinion. In such cases, the exchange of arguments (System 2), will then either reinforce peoples' own preliminary judgment or cause them to arrive at a different judgment: "They are often willing to become convinced of the opposite of their initial intuition" (Musschenga, 2008, p. 135). Additionally, even if people partaking in the discussion behave like a lawyer, that does not mean that they only make up arguments after the fact. After all, the intuitive position they passionately defend may be the result of well-thought-out judgments made in the past. For example, if an individual who grew up in a family in which eating meat was normal made a conscious decision to stop eating meat, they would no longer have to think hard about buying meat every time they go to the supermarket.

Secondly, Musschenga is more optimistic than Haidt about the possibility of adjusting our judgment based on reflecting on new arguments. Most people have the experience of suddenly coming to an entirely different understanding as a result of new perspectives, new information, or new arguments. That thinking process is not being done justice if one claims that only already existing (latent) intuitions and emotions are at work in such a situation.

Thirdly, according to Haidt, System 2, conscious reasoning, is only used in complex situations or when intuitions conflict. However, Haidt appears to ignore new situations for which one has no intuitions to fall back on. Musschenga suspects that situations like this occur on a regular basis, meaning that the statement that System 2 is only rarely needed would not be correct.

Fourthly, although it is true that our reasoning has shortcomings, that people often behave like lawyers, and that they tend to conform to the views of their friends and weigh arguments in a biased way, it must be stated that this is not always possible. As indicated before: people do not always have (strong) intuitions. Additionally, even lawyers will adjust intuitive judgments "if there are too many reasons pleading against it" (Musschenga, 2008, p. 138).

Fifthly, Haidt believes that System 2 will rarely motivate people to behave according to its conclusions. His premise is that only people's emotions (affects) motivate them, and in his view, he allocates those emotions to System 1. Consequently, it follows automatically that System 2 will have little influence on our actions. Yet, even if one goes along with the idea that only our emotions can motivate us, which not everybody does, there are plenty of philosophers who reject such a strict separation between emotions and reasoning (cf. Roeser, 2010). In other words, emotions also play a role in System 2. Additionally, we know from research that people are very attached to their moral identity: they want to see themselves as moral beings and they also want to be known as such by others (cf. Ellemers et al., 2019). This desire is a strong motivator to apply all the resources for moral judgments one has at their disposal, including System 2.

Finally, moral judgment is not always an individual matter: there are many situations in which moral issues have to be dissolved within a group. One such common situation is at work. Different views may exist within a group, or a group may encounter situations with which they are not yet familiar. In such cases, the ability to reason, to provide strong arguments, and to justify certain positions is very valuable (Musschenga, 2009, p. 609). Additionally, transparency and the justification of choices are essential parts of professional life in today's world: professionals must be able to justify their decisions with words that others can understand. In such an environment, actions cannot be justified solely based on intuition. Doing so would be highly problematic because it makes it impossible for another person to judge a particular course of action.

A More Important Role for System 2

Before we move on, an intermediate conclusion is: the role of System 2 seems much bigger than Haidt claims and is also more important in certain situations than Haidt hints at (think about novel situations, group decisions, and disagreements). Additionally, Musschenga shows that System 1 is not always reliable, which is a good reason to ascribe a more substantial role to System 2.

As indicated earlier, our intuitions are partly innate and partly learned. Haidt extensively discusses the evolutionary origins of many intuitions, positing that they must be helpful to humanity as they have, among others, helped humanity get where it is today and have past the 'survival of the fittest' test.[2] However, it should be noted that humanity's evolution took place under very different conditions than the modern, complex society in which those intuitions act as our guide today (cf. Musschenga, 2010; Sauer, 2019). Important features of our innate intuitions include:

- We react primarily to dangers we can see, hear, smell or feel and are less likely to be frightened by dangers we cannot perceive with our own senses.
- The interests of family members outweigh those of strangers.
- Since survival took place on a daily basis, we are fairly short-sighted and have little regard for the future.
- We are strongly conformist: if someone fled it was better to follow them than to stand still.
- We are status sensitive because status is sexy (Giphart & Van Vugt, 2016, cf Flanagan, 2017).

The distinction between System 1 and System 2 has been adopted from the behavioral sciences. In its original context, the model is mainly used to identify different kinds of 'biases'. In essence, the model is used to show how intuitions function as rules of thumb that can systematically misguide individuals (cf Kahneman, 2011; Sunstein, 2005). Based on this, it would seem sensible to not canonize System 1 without question. Yet, this is partly an empirical question: how reliable are peoples' moral intuitions, to what extent is it wise for people to rely on their feelings? In the behavioral sciences, this is tested by determining if System 1 prescribes the best decision. This is done by testing if the same judgment would be reached if System 2 was used. However, when it comes to moral issues it is very difficult to conclude what 'the best decision' is because different ethical theories can justify different outcomes (cf. Brand, 2016; Sunstein, 2005). Therefore, Musschenga (2009) takes a different approach when questioning the reliability of System 1. He does not look at 'the best decision'. Rather, he poses the question whether there are good reasons to trust the wisdom of System 1. Are our intuitions epistemically justified? After all, "Epistemic justification does not require true beliefs. That is, one can have epistemically justified false beliefs" (Liao, 2011).

In order for the intuitions provided by System 1 to be epistemologically justified, System 1 would need to be impartial, i.e. not to make its own interests more important than the interests of others. This is a tall order. As Adam Smith already knew, and as has been confirmed by 'behavioral economics' in many ways, what is close (our own interests) always seems big, and what is far away (the interests of others) seems small. Smith wrote that "it is only by consulting this judge within [our conscience], that we can ever see what relates to ourselves in its proper shape and dimensions; or that we can ever make a proper comparison between our own interests and those

[2] Additionally, evolution has also resulted in System 2. Hence, it is likely that this system is also helpful to humanity (cf. Slovic et al., 2007, p. 1347).

of other people" (Smith, 1759/2009, p. 157). Consequently, System 2 is required to play the role of conscience, of criticizing our intuitions, our moral emotions, and needs to correct them if necessary.

Additionally, System 1 is only reliable if it is not overwhelmed by emotions, something that can't always be ruled out. Although there may be wisdom in emotions, there is a reason why calling someone 'emotional' is generally not a positive qualification. Emotions can get in the way and actually be dysfunctional by functioning as a jammer.

Moreover, you would want System 1 to be insensitive to, for example, the words with which a situation is explained or the circumstances in which a decision must be made. Yet, a variety of studies show that this is not the case (Musschenga, 2009, pp. 604–605). These studies show that our judgment is partly determined by the choice of words or by the order in which options are presented to us. For example, research has been done on people's responses to a hypothetical situation in which they are given a choice between an option in which one person dies and an alternative in which five people die. The results show that whether emphasis was put on the negative outcome (people die) or the positive outcome (people are saved) had significant impact on both the choice people made and their degree of agreement. In a similar vein research has shown that subjects' decision making was different based on whether the decisions were made in a clean room or a dirty room. These factors should be irrelevant to the decision making process but they have been shown to have an impact.

Consequently, we have good reasons not to trust System 1 blindly. System 2 too should be used to critically examine and, if necessary, correct the outcome of System 1. However, the question then becomes how to make the role and importance of System 2 bigger, or whether there may be other ways to enhance the power of System 1.

Methods for Ethics Education

As discussed earlier, System 1, and with it our intuitions and emotions, plays an important role in moral judgments. Yet, System 1 is not unquestionably reliable: System 2 is needed to monitor, test, and, if necessary, correct System 1. Sadly, System 2 is not always reliable either. Fortunately, there are several ways to strengthen trust in both systems.

Strengthening System 2

Classical ethics education focuses on System 2, with a particular emphasis on reasoning using philosophical insights. Although this remains important, Haidt has shown that people primarily think critically in social settings and that the process is prone to developing biases. Consequently, Musschenga looks at de-bias strategies used to enhance the quality of decision making processes and thus use the power of the group to make the individual aware of blind spots and new perspectives (cf. Liao, 2011). Building on Haidt, Mlodinow acknowledges that the mind sometimes behaves like a lawyer but believes that the mind can also behave like a scientist: "Scientists gather evidence, look for regularities, form theories explaining their observations, and test them. Attorneys begin with a conclusion they want to convince others of and then seek evidence that supports it, while also attempting to discredit evidence that doesn't" (Mlodinov, 2013, p. 200). According to Mlodinov, our mind, System 2, is capable of playing both roles: "both a conscious seeker of objective truth and an unconscious, impassioned advocate for what we *want* to believe … As it turns out, the brain is a decent scientist but an absolutely *outstanding* laywer" (Mlodinov, 2013, pp. 200–201). Hence, one way of improving System 2 is putting the scientist to work. This is a metaphor that can be elaborated on using knowledge on the scientific process and the ways scientists operate. A scientist always tries to look at counterarguments, at the arguments that undermine his/her positions. Subsequently, they are willing to revise their position based on the strength of the arguments. Since nothing is foreign to scientists, this is hard work. Therefore, the strength of science lies not in the individual, but in the collective as scientists keep each other on the ball: "It is not so much the critical attitude that individual scientists have taken with respect to their own ideas that has given science the success it has enjoyed as a method for making new discoveries, but more the fact that individual scientists have been highly motivated to demonstrate that hypotheses that are held by some other scientist(s) are false" (Nickerson, 1998, p. 194). Essentially, this view could be used to make a strong argument for the organization of peer review so professionals could discuss moral issues in a systematic way. Practice in moral consultation could be made part of ethics education.

Verweij (2016) also emphasizes the importance of moral case deliberation (MCD) and refers to an "Aristotelian method for MCD that specifically focuses on the reflection on emotions, or rather on the 'rightness' (i.e. the adequateness) of the expression and the reliability of an experienced emotion in a particular situation". Additionally, Verweij reports positive effects on participants as a result of the implementation of MCD: "Participants of MCD sessions in which emotions are addressed report that the quality and the thoroughness of moral inquiry was increased and that they experienced a deeper and more personal learning process" (Verweij, 2016, p. 39). Through the MCD process the reflective skills of System 2 are explicitly used to explore and test the intuitions and emotions of System 1. This does not have to be limited to classroom sessions with power point presentations and instruction cards for real life dilemmas. Such a scenario has the risks "turning dilemmas into 'can-do' challenges that can be solved with a checklist" (Molendijk, 2019, p. 190). Instead, as Molendijk (2019, p. 190) points out, the learning experience is likely to be enhanced by integrating ethics education into field exercises to allow realistic, experience-oriented training.

Another situation in which discussion could be invaluable, is when studying how exemplary figures deal with their emotions. In the Aristotelian tradition, observing others is a valid strategy in critically reflecting on one's own intuitions and emotions:

"If one wants to gain a greater understanding of what a healthy professional life is supposed to look like, one can inquire with those people who he/she intuitively perceives to be knowledgeable" (Sanderse & Kole, 2018, p. 184; our translation). However, to not blindly copy the behavior of role models, one needs to master the art of 'critical and conscious copying'. This is where a conversation with the role model, if possible, could be very valuable because mere observation does not reveal everything. Conversations with role models might provide deeper understanding of why they do what they do.

Strengthen System 1

Although there are several methods to strengthening System 2's reflective skills and relating them to System 1's emotions and intuitions, Musschenga notes that there is not always time to apply System 2. Consequently, he advocates that it is also wise to strengthen the trust in our intuitions, in our feelings, by investing in the reliability of System 1. Since intuitions and emotions are partly learned and partly innate, the learning process in which our intuitions and emotions (partly) take shape is imperative. Here, learning through experience plays an important role. This requires that people get feedback immediately, and that they get feedback from the right people (i.e. not from immoral people or people who don't know their own craft). After all, lack of feedback or unreliable feedback may result in the wrong conclusions being drawn: "You cannot learn from feedback you do not receive and some feedback may simply act to increase confidence in erroneous beliefs" (Musschenga, 2009, p. 606).

Consequently, this is an argument for a structured approach to learning-by-doing. Musschenga refers to the model of Dreyfus and Dreyfus (1991, 2005) that indicates how professionals can become proficient step by step; from beginner, to advanced, to competent and proficient, and eventually to a moral expert:

> It seems that beginners make judgments using strict rules and features, but that with talent and a great deal of involved experience, the beginner develops into an expert who sees intuitively what to do without apply applying rules and making judgments at all. The intellectualist tradition has given an accurate description of the beginner and of the expert facing an unfamiliar situation, but normally an expert does not deliberate. He or she neither reasons nor acts deliberately. He or she simply spontaneously does what has normally worked and, naturally, it normally works. (Dreyfus & Dreyfus, 2005, p. 253)

However, Musschenga has difficulty calling someone a moral expert:

> According to the Dreyfuses, someone who had a normal, successful moral education is a moral expert. What they call moral experts can be better described as ordinary, morally competent persons. While individuals keep being confronted with new situations and new problems, moral development never really ends and needs to continue in adult life. (2009, p. 607)

Although this comment may be justified, the main point of the Dreyfuses still stands: by becoming competent, by learning-by-doing, one could have more confidence in intuitions and emotions. Yet, there is a pitfall: arrogance, indifference and cowardice also come about as a result of practice. Therefore, professionals should preferably practice under the guidance of people with experience on appropriate emotional responses, and thus are able to provide targeted feedback (see above). Consequently, philosophical insight can also play a role: "they figure effectively in the acquisition, formation and maintenance (that is, the education) of subjects' moral intuitions, and make a psychologically real difference to people's moral beliefs. Effective moral reasoning requires nothing more than this" (Sauer, 2012, p. 263).

One way in which professionals could practice under guidance is through the master-apprentice relationship. Although this is a very old form of learning, it certainly is not old-fashioned (Brockmöller, 2008). A master is an expert who has mastered his craft, someone who has achieved superior performance. They have not only done so through knowledge and skills, but also through the values and standards that characterize good professional practice: the master does the right thing the right way. As part of the master-apprentice relationship, the master and the apprentice go to work together and subsequently learn from each other. Since the master is an expert, the apprentice learns from the master. However, since the apprentice asks questions about the practice, the master will become more aware of his own actions and gain new insights. Brockmöller has shown that this process can work, and be mutually beneficial, but the master-apprentice relationship requires a lot of attention. Therefore, she formulated a protocol how to model this relationship.

Conclusion

According to Verweij, more attention should be paid to emotions, or the 'e-word', in military ethics education. Based on the dual-process model this conclusion is justified. Additionally, the model also provides insight on how this increased focus could take shape. Obviously, a change to an increased focus on emotions will not come easily to the military: "the masculine ideal of the warrior hero, and to be in emotional control, does not make it easy to engage in reflection as this could simply be interpreted as being weak or vulnerable" (Van Baarle, 2018, p. 122). Yet, clearly, there is a lot to be gained by doing so: "emotions – regardless of intensity – are inevitable in morally challenging interactions, and influence a serviceman's behavioral responses. Therefore the military would benefit from expanding its attention in the direction of affective processes" (De Graaff, 2016, p. 91). Possibly, semantics could play a role in changing the perspective on emotions. Based on the dual-process model it is clear that emotions and intuitions are closely related. Yet, where emotions are considered taboo, intuitions are associated with expertise: experts perform at a high level within their field based on their intuitions. Expertise is something that most people are willing to strive for. Yet, if people are to actually reach this level, it is a prerequisite that the education program of which they are part of has to have cultivated the means for them to do so. Therefore, emotions can be made part of military ethics education through their inextricable link to intuitions. By addressing intuitions into the learning process in a structured way, by making clear that people could grow from beginners to experts by strengthening their intuitions through education, it is possible to entrench emotions as a component of military ethics education.

References

Brand, C. (Ed.) (2016). *Dual-process theories in moral psychology*. Springer.

Brockmöller, A. (2008). *Knowledge sharing in expert-apprentice relations: Design of a protocol*. University of Groningen.

de Graaff, M. (2016). *Moral forces: Interpreting ethical challenges in military operations*. Universiteit Twente.

Dreyfus, H., & Dreyfus, S. (1991). Towards a phenomenology of ethical expertise. *Human Studies, 14*, 229–250.

Dreyfus, H., & Dreyfus, S. (2005). Expertise in real world contexts. *Organization Studies, 26*(5), 779–792.

Ellemers, N., Van Der Toorn, J., Paunov, Y., & Van Leeuwen, T. (2019). The psychology of morality: A review and analysis of empirical studies published from 1940 through 2017. *Personality and Social Psychology Review, 23*(4), 332–366.

Flanagan, O. (2017). *The geography of morals. Varieties of moral possibility*. Oxford University Press.

Giphart, R., & van Vugt, M. (2016). *Mismatch. Hoe we dagelijks misleid worden door ons oeroude brein*. Podium.

Greene, J., & Haidt, J. (2002). How (and where) does moral judgment work? *Trends in Cognitive Sciences, 6*, 517–523.

Haidt, J. (2001). The emotional dog and its rational tail: A social intuitionist approach to moral judgment. *Psychological Review, 108*, 814–834.

Haidt, J. (2012). *The righteous mind*. Knopf Doubleday Publishing Group.

Haidt, J., & Bjorklund, F. (2008). Social intuitionists answer six questions about moral psychology. In W. Sinnott-Armstrong (Ed.), *Moral psychology, Vol. 2: The cognitive science of morality*. MIT Press.

Haidt, J., & Kesebir, S. (2010). Morality. In S. Fiske et al. (Eds.), *Handbook of social psychology*. Wiley.

Kahneman, D. (2011). *Thinking, fast and slow*. Allen Lane.

Karssing, E. (2021). Uit de boekenkast van de bedrijfsethiek 82. *Tijdschrift voor compliance, 21*, 232–240.

Liao, S. (2011). Bias and reasoning: Haidt's theory of moral judgment. In T. Brooks (Ed.), *New waves in ethics*. Palgrave Macmillan.

Mlodinov, L. (2013). *Subliminal*. Random House.

Molendijk, T. (2019). *Soldiers in conflict. Moral injury, political practices and public perceptions*. Radboud University.

Musschenga, M. (2008). Moral judgement and moral reasoning. A critique of Jonathan Haidt. In M. Düwell et al. (Eds.), *The contingent nature of life*. Springer.

Musschenga, M. (2009). Moral intuitions, moral expertise and moral reasoning. *Journal of Philosophy of Education, 43*(4), 597–613.

Musschenga, M. (2010). The epistemic value of intuitive moral judgements. *Philosophical Explorations, 13*(2), 113–128.

Musschenga, M. (2011). The relevance of conscious moral reasoning. In D. de Ruyter & S. Miedema (Eds.), *Moral education and development*. Sense.

Nickerson, R. (1998). Confirmation bias: A ubiquitous phenomenon in many guises. *Review of General Psychology, 2*(2), 175–220.

Roeser, S. (2010). Intuitions, emotions and gut reactions in decisions about Risks. *Journal of Risk Research, 13*(2), 175–190.
Sanderse, W., & Kole, J. (2018). Deugdzaam worden. Praktische werkvormen, In W. Sanderse & J. Kole (Eds.), *Karakter. Deugden voor professionals.* ISVW.
Sauer, H. (2012). Educated intuitions. Automaticity and rationality in moral judgement. *Philosophical Explorations, 15*(3), 255–275.
Sauer, H. (2019). *Moral thinking, fast and slow.* Routledge.
Sinnott-Armstrong, W., Young, L., & Cushman, F. (2010). Moral intuitions. In J. Doris (Ed.), *The moral psychology handbook.* Oxford University Press.
Slovic, P., Finucane, M. L., Peters, E., & MacGregor, D. G. (2007). The affect heuristic. *European Journal of Operational Research, 177*, 1333–1352.
Smith, A. (1759/2009). *The theory of moral sentiments.* Penguin.
Sunstein, C. (2005). Moral heuristics. *Behavioral and Brain Sciences, 28*, 531–542.
van Baarle, E. (2018). *Ethics education in the military: Fostering reflective practice and moral competence.* Vrije Universiteit.
Verweij, D. (2016). Why address the 'E'-Word in military ethics education? The role of emotions in moral judgment and decision-making. In T. Elssner & J. Reinhold (Eds.), *Didactics of military ethics. From theory to practice.* Brill/Nijhoff.

Prof. Edgar Karssing is professor of Philosophy, professional ethics and integrity management at Nyenrode Business Universiteit. He received his Masters degree from Rijksuniversiteit Groningen in both Economics and Philosophy. He has completed a PhD thesis on integrity in theory and practice.

[illegible]

[illegible]

The Dutch Approach to Ethics: Integrity Management in the Military

Miriam Carla de Graaff and Claire Zalm

> A commercial party invites a commanding officer for an exclusive dinner in a gourmet restaurant: the tender has been awarded. Is it wise to accept the offer? A military instructor is found to be rather intimidating and imposes higher demands on cadets than is required in the curriculum. Is this acceptable Bildung?

The above described situations are examples of daily ethical issues that are hardly ever considered ethical issues or dilemmas and fail to receive much attention in military ethics training. However, just like the more tragic ethical dilemmas such as, 'whether to shoot or not', they are not as easy to answer as might seem at first glance. At a personal level, individuals may be tempted to act in a certain way, regretting their choices afterwards. Or, they may feel confused about the right thing to do. Such daily dilemmas are at the centre of ethics management in the Dutch military: integrity management, which pertains to the professional performance of duties. Integrity management in the context of the military relates to due care, to doing justice and decency and to reliability with respect to the citizens of the Netherlands, the countries in which the Netherlands Defence organisation is active, and also with

---

The authors of this article were both employed with the Netherlands Defence Centre of Expertise for Integrity (COID); Claire Zalm until 2018, Miriam de Graaff until 2019. Claire Zalm is currently employed as the Director of Security, Crisis Management and Integrity at the Ministry of Foreign Affairs. Miriam de Graaff is currently employed as Head of the Joint Centre of Expertise for Communication and Engagement within the Royal Netherlands Army. The authors wrote this contribution in a private capacity and the viewpoints expressed do not necessarily reflect the position and/or policy of the Netherlands Ministry of Defence. The authors would like to express their gratitude to Tine Molendijk Ph.D. and Professor Dr. Eric-Hans Kramer for their feedback on previous drafts of this chapter.

M. C. de Graaff (✉)
Royal Netherlands Armed Forces Joint Centre of Expertise Communication & Engagement, Apeldoorn, The Netherlands
e-mail: MC.d.Graaff@mindef.nl

C. Zalm
Ministry of Foreign Affairs, Department of Security, Crisis Coordination & Integrity, The Hague, The Netherlands

E.-H. Kramer and T. Molendijk (eds.), *Violence in Extreme Conditions*,
https://doi.org/10.1007/978-3-031-16119-3_11

respect to its own personnel by providing them protection within the framework of an employer's responsibility. In this contribution, we outline the pillars of the integrity management system of the Netherlands armed forces. The purpose of this contribution is to provide some background on integrity management for public managers who currently work in a high-stakes environment such as the military. It is not our intent to claim that the Dutch approach is a one-size-fits-all solution; but rather our purpose is to start a discussion within the field of military ethics on how integrity management can be carried out in environments that are highly demanding, using the Dutch military as an example.

Managing Ethics, or Integrity Management

Ethical transgressions have a large impact on an organisation's output, financial position and its employees, for example when due to counterproductive work behaviour, fraud or *administrative evil* (Kolthoff, 2016). However, unethical behaviour can have enormous societal impact as well due to state crime and human rights violations (Kolthoff, 2016). This is especially true in organisations that operate in the public sector, as violations by these organisations can lead to environmental hazards, health risks and may affect the personal lives and wellbeing of all individuals and parties involved (Hoekstra & Heres, 2016; Kolthoff, 2016). The importance of integrity management therefore lies not only in the contribution to positive organisational outcomes, but also in preventing the negative effects that a lack of public integrity may result in. It is for that reason that integrity is considered the cornerstone of *good governance* in the public sector of the Netherlands (Hoekstra & Heres, 2016). This emphasis on ethics, integrity management and integrity policy should also be present in military practice (cf. De Graaff, Schut, et al., 2017).

In academic literature, the terms *integrity* and *ethics* are often used interchangeably and as synonyms. However, in practice one often thinks of rule-following *compliance* when using the term integrity, whereas ethics often refers to a broader concept related to major moral choices, such as considerations as to whether or not participation in a war is to be considered *just*. In this chapter, we address the obligation of the military organisation to be a good employer to its employees and the positive pay off this has for all parties its personnel engages with. In many organisations, two types of management strategies aiming to promote morally responsible behaviour can be distinguished (Paine, 1994). The first is a rule-based and top down *compliance approach*, focusing on the prevention of misconduct and transgressions of laid down rules and procedures. The second is referred to as the *integrity approach*, which is value-based and bottom-up focusing on supporting individuals in making morally responsible decisions (Paine, 1994). Within the Netherlands armed forces, both approaches are recognised and both contribute to the organisation's aim for good governance.

In this chapter, we consider integrity management to be a three-layered framework consisting of three distinct layers wherein both of the mentioned strategies can be implemented. These layers are individual competencies, ethical climate and organisational design; integrity management based on these three layers results in decisions being made professionally, prudently and in such a way as to do justice to all parties involved.

The individual level (layer one). Activities in this layer concern stimulating the ability and the desire to make morally prudent choices, even if relevant legislation is unclear, lacking or not applicable. This individual agency perspective is also introduced and further deepened by Desiree Verweij as individual *moral competence*[1] (Verweij, 2007) and is in line with Richardson, Verweij and Winslow's perspective of *moral fitness* (2004). In most cases, stimulating individual morally responsible behaviour starts in a classroom or group training. This type of didactics works both ways in establishing individual competence through moral reflection as well as in mores or group culture (Van Baarle et al., 2015). Mores entail the second layer in integrity management. Individual awareness of the moral dimensions of any situation is activated when others share their dilemmas and ideas on how to solve an issue (Van Baarle et al., 2015). Also, being in a training setting together stimulates the onset of discourse, meaning in these settings all participants are forced to verbalise their moral intuitions and emotions. As such, group exercises and individual moral reflection helps in communicating about moral issues (Van Baarle et al., 2015).

However, the effectiveness of these ethics programmes in organisations is not always easy (or even possible) to identify. It is often also subject to debate. For example, Wang and Calvano (2015) showed that gender differences influence the effectiveness of certain aspects of ethics programmes. Weaver (2001) argues that the effectiveness of these ethics programmes may well be culturally undermined. On the other hand, in the Dutch military the results of ethics programmes for military instructors seem promising (Van Baarle et al., 2017), as do the results of Canadian battlefield ethics training (Thompson & Jetly, 2014).

Group level/ethical climate (layer two). The second layer addresses the way people interact with one another. Examples of this second layer include informal standards and mores about employee voice and *esprit de corps*. A relevant aspect of team culture in this respect is the influence of role models and beliefs about what virtues the organisation or team stand for. In military training, a great deal of focus lies

[1] Moral competence comprises six elements (based on Verweij, 2007):

(a) Moral awareness—recognising the moral dimension of a situation.
(b) Self-reflection—being aware of one's personal standards and values and possible bias about a situation.
(c) Moral understanding—being able to formulate various courses of action and their consequences.
(d) Moral opinion forming—being able to and willing to make a decision and act accordingly.
(e) Responsibility and communication—being able and willing to communicate the reasons and considerations underlying a choice made to others. Taking responsibility.
(f) Moral resilience—being able to cope with the tragic consequences of choices made.

on the transfer of an *esprit de corps* in terms of shared virtues such as loyalty and obedience. However, the traditional military virtues that lie within this *esprit de corps* lead to a strong focus on one's own group, disregarding other parties involved and other perspectives in the situation at hand (Verweij, 2013), which may in turn result in ethical failure by social psychological mechanisms such as groupthink (cf. De Graaff, De Vries, et al., 2017). It is therefore relevant to educate servicemen and women in recognising these mechanisms and their negative effects.

Both during military operations and during general peacetime operations, incidents that cannot pass muster do take place. Commanders are responsible for properly responding to any reports submitted by anyone in their chain of command. The core principle of the Dutch integrity policy is that members of staff call each other to account in the case of unacceptable behaviour. The main focus is on doing this timely and respectfully in order to prevent situations and behaviour from escalating. This is referred to as *employee voice*. A study conducted in the Netherlands armed forces on prosocial voice (i.e. attempting to improve the situation by addressing the behaviour of co-workers by expressing one's own opinions and feelings) showed that when individuals consider it to be *normal* in their working environment to speak up and confront co-workers, they are more inclined to do so regardless of the behaviours of others and what they actually see that others are doing in terms of voice (Hilverda et al., 2018).

The structural/design level (layer three). The third layer is made up of the role played by the organisation to encourage and facilitate its personnel to perform their work in a morally prudent fashion. This concerns the formal structure of the organisation and involves the organisation being aware of the vulnerable position of its personnel, of high-risk processes and of legislative developments (De Graaff, Schut, et al., 2017; De Graaff & Van den Berg, 2010).

Some years ago, a risk analysis was performed on the Dutch officers' training programme (Governance and Integrity, 2013). One of the conclusions of this analysis was that the final assessment of cadets could be made more objective and would benefit from further standardisation, so as to guarantee a more equal treatment. This resulted in a multi-disciplinary project being launched to further professionalise the training process. The project team worked on, *inter alia*, making the instructors aware of their crucial and, at the same time, vulnerable position in the training process, on embedding integrity into the instructors' training courses, on evaluating the cadet assessment process and on reformulating the course requirements. The cadets and their instructors themselves were also involved in the project. At the same time, stock was taken of the way ethics and integrity were taught in the various career training programmes and to what extent this was in line with the duties cadets are expected to perform upon finishing the programme concerned.

Doing justice to all parties involved is the core principle of the integrity policy of the Netherlands Defence organisation (Secretary-General, no date). This is formulated as follows: *To treat each other and others with respect, taking account of the rights, interests and wishes of all parties involved*. This does not mean that everyone will always be *happy* with the choices that are made. What it does mean is that all choices can be explained and that a decision is not based on a single point of view. Such also becomes apparent from the reference made in the organisation's integrity policy to the term *respect*. The Latin verb *respectare* has multiple meanings, including *looking after others*. In other words, when acting from this perspective, whenever there is interaction with others, the MOD wants to look after and make allowances for the persons involved. In the Dutch Defence Code of Conduct, this policy has been translated into cornerstones that are relevant to all employees and that in the main relate to manners and conduct. The (value driven) cornerstones are: commitment (in Dutch: *verbondenheid*), safety (in Dutch: *veiligheid*), trustworthiness (in Dutch: *vertrouwen*) and responsibility (in Dutch: *verantwoordelijkheid*) (Ministry of Defence, 2018).

Institutionalising Integrity Management in the Dutch Military

The Netherlands armed forces organise both preventive activities and activities based on violations that may be expected due certain vulnerabilities and risks in the working procedures and parties involved. Within the armed forces, several departments cooperate in initiatives providing support to the Defence organisation and its staff by performing preventive activities in all three of the described layers. For example, they cooperate at strengthening moral competence by providing so-called dilemma training sessions (layer one), providing insight into the level of the ethical climate within a unit by conducting research into the culture (layer two), and charting vulnerable organisational structures and working processes by performing risk analyses (layer three). Departments that play a role in these prevention activities are, *inter alia*, the Defence Centre of Expertise for Integrity (COID),[2] the School for Peace Operations (SVV) and the Defence Centre of Expertise for Leadership Development (ECLD).

Scientific research has shown that Dutch military personnel face various dilemmas during military operations, for instance in the context of experiencing cultural differences (cf. De Graaff et al., 2016; Schut & Moelker, 2015). In many countries, ethics training is carried out in relation to military operations and deployment, such as Canada (Warner et al., 2011), the Netherlands (De Graaff, Schut, et al., 2017), Switzerland and the United States (Williams, 2010). In the Netherlands, for instance, it is customary to prepare ship's crews, prior to a long-term posting at sea, for confrontations with possible temptations and possible ethical dilemmas; temptations

[2] The COID is a centre of expertise that supports the Defence organisation to do justice to all parties involved and to do so in a respectful manner. It provides such support by, *inter alia*, providing advice, training courses and workshops, and by performing investigations and analyses, both preventatively and following suspected violations. The COID is internationally considered an example for other armed forces to follow, as is recognised in the 2016 van der Steenhoven report on integrity within the Netherlands Defence organisation (Van der Steenhoven & Aalbersberg, 2016).

that come with the posting and run counter to organisational rules and standards. Examples are having to deal with seized contraband such as drugs; ribbing that turns to bullying due to being cooped up in close quarters for a prolonged period of time; or cultural differences resulting in undesirable behaviour. However, when it comes to the daily issues as described in the introduction, only a few programmes focus on this aspect of integrity management. We argue that these efforts should be intensified. Such educational and training activities may not only result in positive organisational results (i.e. fewer incidents and scandals), but may also have a positive effect on employee well-being (Thompson & Jetly, 2014) by acknowledging how an individual is influenced by these situations in terms of moral emotions (De Graaff et al., 2016; Schut et al., 2015) and maybe even moral injury (Molendijk, 2018).

When situations do not only cross the line of what employees individually consider to be acceptable behaviour, but violate organisational boundaries as well (e.g. misconduct and fraud), employees ought to report such behaviour to their commander, who will take further action. In (potentially) harmful and complex cases, employees have the possibility of *whistleblowing*, meaning they consult a third party before stakeholders are involved in resolving the complaint. Should the employee not or not yet be sure of which action to take, they may contact a confidential advisor (in Dutch: *Vertrouwenspersoon*). Such advisors provide emotional support, are familiar with the procedures and are able to suggest alternative, informal solutions. The Netherlands Defence organisation includes a network of about 600 confidential advisors who perform their advisory work in an ancillary position. In addition, all personnel may consult and discuss personal dilemmas with Military Chaplaincy personnel. This is a network of officials contributing to the (spiritual) welfare of military personnel and other Defence staff and to the morality of the armed forces as a whole from a Jewish, Roman Catholic, Protestant, Humanist, Hindu or Islamic tradition, education and background.

The Netherlands armed forces also have a long tradition of the so-called 'Non-Commissioned Officers Chain', the senior non-commissioned officer serving as an assistant to the commander in case of incidents. Because of the role and position, the senior non-commissioned officer is the obvious person in the chain of command to call attention to problems and think of solutions. This role and position also allows the non-commissioned officer serving as assistant to the commander to broach issues at the right level, mediate where necessary or settle a case with a customised solution on behalf of the commander.

In addition to reporting an incident up the chain of command, staff may also report to the Integrity Reporting Office (in Dutch: *Meldpunt*) if they feel they are not heard by their commanding officers for any reason, or are afraid to report to them. For commanding officers themselves, the previously mentioned COID is an internal department that focuses on providing support to those commanding officers in dealing with suspected violations meticulously; this entails, for instance, rendering advice on how to deal with reports and what to communicate, on the suitability of conducting an investigation, and, when necessary, on providing investigators. Should an investigation be initiated, the legal department and HR also often play a part.

At times, personnel are unable to resolve their problems within the organisation and with the support of the existing systems. In such cases, they may contact the Inspector-General of the Armed Forces (*Inspecteur-Generaal Krijgsmacht*, IGK), an official who acts outside of the formal organisational structure of the Ministry and who directly reports to the Minister of Defence. Each individual military or civilian member of staff or their family may contact the IGK. The IGK listens to their concerns and tries to bring the parties together to work on a solution, if necessary. Obviously, next to the possibility of consulting the IGK, the previously mentioned office for whistleblowing is a possibility at hand.

Conclusion

To summarise, the Netherlands armed forces are equipped with a plethora of departments, officials and initiatives to increase moral fitness within the organisation. Yet despite all of these measures (drawn up in writing), ethics officials (from coaches, integrity advisors, analysts and instructors) ought to stay connected and keep their 'boots on the ground' in order for integrity management to be both practical and to be put into practice. Put into practice not only by these officials, but by all personnel. Doing so will ensure that integrity management is not reduced to mere window-dressing, but in fact receives the importance it deserves and is carried into effect by all.

We argue that for coming to a morally responsible solution in day-to-day ethical dilemmas as described in the introduction, similar activities are necessary in integrity management as for answering the broader ethical questions regarding just war principles or tragic dilemmas in, for example, encountering hostile non-combatants or child soldiers. Therefore, we believe integrity management and day-to-day (peacetime) ethical issues should be more integrated into military ethics training, discussions and considerations. Such issues ought not to be disregarded and neglected, as they may result in organisational problems, individual issues and negative emotions in the long term.

References

De Graaff, M. C., De Vries, P. W., Van Bijlevelt, W. J., & Giebels, E. (2017). Ethical leadership in the military, the gap between theory and practice in ethics education. In P. J. Olsthoorn (Ed.), *Ethics in military leadership* (pp. 56–85). Brill.

De Graaff, M. C., Schut, M., Verweij, D. E. M., Vermetten, E., & Giebels, E. (2016). Emotional reactions and moral judgment: The effects of morally challenging interactions in military operations. *Ethics & Behavior, 26*(1), 14–31.

De Graaff, M. C., Schut, M., & Zalm, C. E. (2017). *Morele fitheid in militaire inzet, de noodzaak van aandacht voor integriteit ook in het 'echte' werk.* In N. Boonstra, et al. (Eds.), *Jaarboek Compliance 2017.* Nederlands Compliance Instituut.

De Graaff, M. C., & Van den Berg, C. E. M. (2010). Moral professionalism within the Royal Netherlands Armed Forces. In J. Stouffer & S. Seiler (Eds.), *Military ethics, international perspectives* (pp. 1–24). Canadian Defence Academy Press.

Governance & Integrity. (2013). *Integriteit bij de opleiding en vorming van adelborsten en cadetten aan de Nederlandse Defensie Academie. Duiding van risico's, perspectief of verbetering.* Ministerie van Defensie.

Hilverda, F., van Gils, R., & De Graaff, M. C. (2018). Confronting co-workers: Role models, attitudes, expectations, and perceived behavioral control as predictors of employee voice in the military. *Frontiers in Psychology, 9*, 2515.

Hoekstra, A., & Heres, L. (2016). Ethical probity in the public service. In A. Farazmand (Eds.), *Global encyclopedia of public administration, public Policy and governance.* Springer. https://doi.org/10.1007/978-3-319-31816-5_922-1

Kolthoff, E. (2016). Integrity violations, white-collar crime, and violations of human rights: Revealing the connection. *Public Integrity, 18*(4), 396–418.

Ministerie van Defensie. (2018). *Gedragscode.* Ministerie van Defensie. Retrieved online: April 16, 2022. https://www.defensie.nl/downloads/publicaties/2018/12/04/gedragscode-defensie

Molendijk, T. (2018). Toward an interdisciplinary conceptualization of moral injury: From unequivocal guilt and anger to moral conflict and disorientation. *New Ideas in Psychology, 51*, 1–8.

Paine, L. S. (1994). Managing for organizational integrity. *Harvard Business Review, 72*(2), 106–117.

Richardson, R., Verweij, D., & Winslow, D. (2004). Moral fitness for peace operations. *Journal of Political and Military Sociology, 32*(1), 99–113.

Schut, M., De Graaff, M. C., & Verweij, D. (2015). Moral emotions during military deployments of Dutch forces: A qualitative study on moral emotions in intercultural interactions. *Armed Forces & Society, 41*(4), 616–638.

Schut, M., & Moelker, R. (2015). Respectful agents: Between the Scylla and Charybdis of cultural and moral incapacity. *Journal of Military Ethics, 14*, 232–246.

Secretary General. (no date). *Instruction of the Secretary General 984—Implementation of the Integrity Policy.* Ministerie van Defensie.

Thompson, M. M., & Jetly, R. (2014). Battlefield ethics training: Integrating ethical scenarios in high-intensity military field exercises. *European Journal of Psychotraumatology, 5*(1), 23668.

Van Baarle, E., Bosch, J., Widdershoven, G., Verweij, D., & Molewijk, B. (2015). Moral dilemmas in a military context: A case study of a train the trainer course on military ethics. *Journal of Moral Education, 44*(4), 457–478.

Van Baarle, E., Hartman, L., Verweij, D., Molewijk, B., & Widdershoven, G. (2017). What sticks? The evaluation of a train-the-trainer course in military ethics and its perceived outcomes. *Journal of Military Ethics, 16*(1–2), 56–77.

Van der Steenhoven, K., & Aalbersberg, M. (2016). *Moral Fitness, een onderzoek naar de integriteit van de defensieorganisatie ten aanzien van de omgang met commerciële partijen op het gebied van IT.* ABD TOP Consult.

Verweij, D. E. M. (2007). Morele professionaliteit in de militaire praktijk. In J. Kole & D. De Ruyter (Eds.), *Werkzame Idealen, ethische reflecties op professionaliteit* (pp. 126–138). Koninklijke van Gorcum.

Verweij, D. E. M. (2013). Responsibility as the 'ability to respond' adequately? In H. Amersfoort, R. Moelker, & J. Soeters (Eds.), *Moral responsibility & military effectiveness* (pp. 11–32). T.M.C. Asser Press.

Wang, L. C., & Calvano, L. (2015). Is business ethics education effective? An analysis of gender, personal ethical perspectives, and moral judgment. *Journal of Business Ethics, 126*(4), 591–602.

Warner, C. H., Appenzeller, G. N., Mobbs, A., Parker, J. R., Warner, C. M., Grieger, T., & Hoge, C. W. (2011). Effectiveness of battlefield-ethics training during combat deployment: A programme assessment. *The Lancet, 378*(9794), 915–924.

Weaver, G. R. (2001). Ethics programs in global businesses: Culture's role in managing ethics. *Journal of Business Ethics, 30*(1), 3–15.
Williams, K. R. (2010). An assessment of moral and character education in initial entry training (IET). *Journal of Military Ethics, 9*(1), 41–56.

Dr. Miriam Carla de Graaff is head of the Centre of Expertise for Communication & Engagement within the Royal Netherlands Armed Forces. She obtained a doctorate in Psychology at the University of Twente, in cooperation with the Netherlands Defence Academy. Miriam received a Masters degree (MSc) in Communicationstudies (Cum Laude) and Psychology from the University of Twente. In 2015 she was nominated for the Young Talent Award of the Dutch Women in Leadership election.

Claire Zalm LLM, MPA is Director Security, Crisis management & Integrity of the Ministry of Foreign Affairs. Her Departement is also responsable for stimulating a safe working environment within the MFA and combatting sexual exploitation, abuse and harassment within organisations that recieve funding.

Epilogue

‘Moresfare’ and the Resilience Paradox: Ethics as the Terra Incognita of Hybrid Warfare and Its Challenges

Desiree Verweij

The term ‘hybrid warfare’ is on the lips of many defence specialists worldwide, thus also in Europe. As one can read on its website, the European Community of Interest on Strategy and Defence (part of the European Centre of Excellence for Countering Hybrid Threats) ‘aims at uncovering the essence and nature of hybrid warfare, as well as the logic and pattern of hybrid strategies in order to develop a solid analytical framework as a basis for the assessment of current and future hybrid warfare situations and their practical implications’ (https://www.hybridcoe.fi/coi-strategy-and-defence/).

However, what remains largely unexposed in this endeavour is the connection between violence, power and law and the crucial role of the psychological and ethical dimension in both hybrid conflicts and in the countering thereof. This chapter addresses all these aspects in their interconnectedness, starting with Freud’s basic psychoanalytical insights on war (section “Violence, Power and Law and the Role of Emotional Bonds”), which reveal the connection between violence, power and law and the role emotions play on a political level. This broadened view on war is followed by a discussion of the term ‘hybrid warfare’ in section “Hybrid Warfare”. To better understand this form of warfare and its implications, the term will be contrasted with Kaldor’s notion of new wars, hybrid peace and human security. Kaldor’s ethicopolitical approach helps us to see the importance of a broader focus. It therefore helps us to see the way in which, in hybrid warfare, values and the emotions connected to these values are played upon, influenced, questioned and moulded; in short, used and misused. This way of using and misusing values and norms is what I have called ‘moresfare’ (see Verweij, 2019a). As explained in section “Moresfare and the Dynamic Between Ethics and Law”, the term ‘moresfare’ is derived from the more familiar term ‘lawfare’ (Dunlap, 2009) as a form of hybrid warfare, and can thus be

D. Verweij (✉)
Faculty of Military Sciences, Netherlands Defence Academy, Breda, The Netherlands
e-mail: dverweij@xs4all.nl

CICAM, Radboud University, Nijmegen, The Netherlands

E.-H. Kramer and T. Molendijk (eds.), *Violence in Extreme Conditions*,
https://doi.org/10.1007/978-3-031-16119-3_12

understood as 'the use or misuse of values and norms as a weapon of war or as a way to achieve a political objective' (see also section "Moresfare and the Dynamic Between Ethics and Law"). Moresfare points to the manipulation of values and norms and the emotions associated with them. In this article, the concept of moresfare illustrates the interaction between ethics and law in the context of hybrid warfare and therefore contributes to a sound understanding of hybrid conflicts and of ways to counter them. As this chapter will show, this starts with insight into a systemic ethical approach, and thus with a keen eye for the values at stake in the system as a whole. This means that the values and connected interests of all stakeholders are taken into account, as well as the long-term effects of actions taken. If this systemic ethical approach is absent, countering hybrid warfare and, in line with that, strengthening the resilience of a community are doomed to fail. This danger of moral myopia present in lawfare and the detrimental consequences thereof are discussed in section "The Danger of Moral Myopia in Lawfare". If this moral myopia, which implies a blind spot for the system as a whole, can also be detected in the use of the concept of resilience, problems will accumulate, since both the EU and NATO call for resilience as a way to counter hybrid warfare (section "Countering Hybrid Threats: Resilience as a Panacea?").

Violence, Power and Law and the Role of Emotional Bonds

The raison d'être of the armed forces is directly related to the Weberian notion of the state's legitimate monopoly on violence. This legitimate use of force is entrusted by the state to the armed forces with regard to 'enemies' from outside and to the police with regard to 'enemies' from within the state. The entrusted use of legitimate force underlines the connectedness of power, violence and law, as Freud already pointed out in his discussion with Einstein in 1933 in *Warum Krieg?* This booklet, or rather the exchange of letters, consists of Freud's answer to Einstein's question whether there is 'any way of delivering mankind from the menace of war' (Einstein & Freud, 2018, p. 31). Freud starts by explaining how violence paved (and paves) the way for power and law. The power of the strongest is supported by 'brutal or intelligent use of force' (op. cit., p. 33) and by working together and closing ranks; the power of the strongest alliance determines the law. Law is thus rooted in the power of a community and the violence on which this power is originally founded will turn against any individual or group that opposes this power and its law (op. cit., p. 34). Freud emphasizes that it is a fallacy not to consider the fact that law was originally based on violence and cannot do without the support of coercion and, if necessary, violence (op. cit., p. 40). In the context of this chapter, it is important to note that there is a crucial condition attached to the transition from violence to law. Freud calls this a 'psychological condition' (op. cit., p. 34). In order for a community to exist in a sustainable way, emotional bonds (such as feelings of solidarity) are needed; these emotional bonds hold a community together (op. cit., p. 35). However, these feelings of connectedness do not come naturally. Every community contains unequal power relations, and these form the basis of tensions that may be resolved by adjusting the law, or, as a last resort, by the use of force by those who hold power (op. cit., p. 36). Thus, according to Freud, a community is held together by two factors: the emotional bonds of its members and coercion through violence (op. cit., p. 39). Emotional bonds in the sense of solidarity and connectedness therefore have a political dimension. In continuation of his discussion of the relation between violence, power and law, in *Warum Krieg* Freud addresses the emotions related to two basic human instincts that are typified as Eros (referring to the drive or instinct of life, love and, in a broader sense, the ability to connect to and identify with other people) and Thanatos (referring to the opposite drive or instinct of death, destruction and aggression). Both fundamental instincts or drives coexist in each person, alternating between conflict and harmony in their struggle against each other, a struggle that takes place in and between people. In *Warum Krieg*, Freud discusses the way war appeals to these fundamental instincts, resulting in hate and aggression. As Freud indicates, these destructive tendencies cannot be abolished, but they can be 'bent' (op. cit., p. 44) in such a way that they do not end in war. The best way to do this is to call on Eros for help, since Eros is the counterpart of Thanatos and thus of hate and aggression (op. cit., p. 44). Eros creates the constructive emotional bonds—Freud mentions solidarity and identification (op. cit. p., 45)—between people that forestall destructive conflicts.

What Freud and, in his wake, many other psychoanalysts (and authors) make clear is the important role emotions play, not just with regard to individual issues but also in political contexts. Inspired by Freud, Alford (1998), for instance, analyses the role 'hate' can play in the political arena in terms of its ability to create communities. In this context, Moisi's book on the geopolitics of emotion (Moisi, 2010) can also be mentioned. No matter how convincing the rationalization of political motives may seem, emotions and the values associated with them lie at the basis of every war, including hybrid wars, as many current conflicts are referred to.

Hybrid Warfare

As many authors have indicated, hybrid warfare is not new. However, the fact that the novelty of a concept is contested is not new either. As soon as a new concept is introduced, authors rush to come up with examples from the past that undermine the alleged novelty. This is also the case with the concept of hybrid warfare. On the basis of the often-cited definition by Hoffman (2007), many authors have pointed out hybridity in warfare in ancient Rome, in the Napoleonic era and in the Vietnam War (McCulloh & Johnson, 2013). Even the Trojan Horse is seen as one of the first examples of hybrid warfare (Stoltenberg, 2015). My aim is neither to question nor to underline the novelty of the concept. I do, however, want to point out important similarities between the concept of hybrid warfare and the concept of 'new wars' as introduced by Kaldor (2007).

There are hardly any references to Kaldor in the literature on hybrid warfare, yet her definition of 'new wars' largely corresponds to the often-cited definition of hybrid war used by Hoffman. He describes hybrid wars as a range of different modes of warfare, including conventional capabilities, irregular tactics and formations, terrorist acts, including indiscriminate violence and coercion, and criminal disorder (Hoffman, 2007, p. 14). Building on Hoffman's definition, many authors use similar descriptions. These definitions are in accordance with crucial aspects of Kaldor's definition of new wars as a mix of traditional warfare, large-scale violations of human rights and organized crime (Kaldor, 2007, p. 12). The pre-conflict phases and the post-conflict phases of new wars are very similar, and new wars have the tendency to spread through criminal networks, refugee flows and ideologies (Kaldor, 2007, p. 71). Kaldor calls for a different approach to successfully counter these new wars; an approach in which human rights supersede the rights of states and state power holders in the sense that the focus should be on the security of individuals and communities and not primarily on the security and the interests of the state and its power holders. This so-called human security approach, in which a connection is made between security and development, is closely related to the UN 'responsibility to protect' principle, according to which the security of people comes first and the principle of state sovereignty has to give way to the protection of people who are suffering inhumanely and are in need of basic security.

It seems that Kaldor's ideas on 'new wars' and her reports on 'human security' (respectively the 'Barcelona Report', 2004, and the 'Berlin Report', 2016) can make an important contribution to discussions regarding the phenomenon of hybridity in present-day conflicts and wars. This holds particularly for the last report of the Human Security Study Group entitled *From Hybrid Peace to Human Security: Rethinking EU Strategy towards Conflict*, which proposes a 'second generation human security approach' for the twenty-first century. What these reports make clear is that there are crucial differences between Hoffman (and authors of a kindred spirit) and Kaldor in their evaluation of present-day wars and conflicts. Kaldor's broader focus on 'human security' aims to counter the underlying causes of insecurity and human needs in detrimental living conditions; conditions that lead to structural poverty, insecurity and conflict. This is a completely different approach than the one Hoffman takes. Without taking the roots of conflicts into account, he primarily perceives present-day threats as designed to target Western societies and specifically 'US vulnerabilities' (Hoffman, 2007, p. 14). The answer both authors formulate with regard to hybrid or new wars also differs. Kaldor argues for a human security force, a mix of military personnel and civilian experts that is capable of addressing problems in a bottom-up approach. She thus starts with the needs voiced by the people living in the detrimental and sometimes destructive living conditions that lead to insecurity and conflict. Contrary to this, Hoffman argues for hybrid warriors 'capable of [...] winning on any type of battle space [...]' (Hoffman, 2007, p. 50). Yet, as Kaldor (2008) maintains in her critique on specific just war principles, the focus should not only be on winning the war or defeating enemies. It should also be on creating 'just peace', which implies protection and stabilization and a situation in which legitimacy is related to a growing body of human rights rather than to the rights of states (Kaldor, 2008). The focus is thus not on the other as an enemy or adversary, as Hoffman maintains, but on the people suffering from conflict and therefore on the other as a person in need of peace and security (or to use Roosevelts words, in need of freedom from fear and freedom from want). This difference in their assessment of present-day conflicts is caused by the fact that Kaldor includes a systemic ethical dimension in her analysis and thus addresses the values that are at stake in the (global sociopolitical) system as a whole, whereas Hoffman primarily focuses on the US (and Western societies) as a victim of aggression.

It is important to note that present-day hybrid wars or new wars lean heavily on globalization processes and technological developments. In fact, these processes and developments have made hybrid wars possible. In that sense, the concept of hybrid war cannot be taken out of the broader context in which the phenomenon occurs. Indeed, in order to actually understand this phenomenon and its consequences, insight into its context is crucial, and this context is coloured by geopolitical, social, cultural and moral aspects. The moral aspects in particular are often overlooked, which raise some questions, since values and the emotions related to these values are regularly misused and manipulated in hybrid warfare.

Moresfare and the Dynamic Between Ethics and Law

The Latin 'mores' refers to the traditional customs and ways of behaving that are typical of a community (or part of a community). It thus refers to the underlying norms and values on which customs and behaviour are based. The concept 'moresfare' that I introduced in a paper in 2019 (Verweij, 2019a, b) is closely related to the concept 'lawfare' as discussed by several authors but introduced by Dunlap in 2001. He described lawfare as the 'use of law as a means to realizing a military objective' and thus as a weapon of war (Dunlap, 2001, p. 4). In an article written eight years later, Dunlap uses a broader definition: 'the strategy of using or misusing law as a substitute for traditional military means to achieve an operational objective' (2009, p. 35). In a similar way, I describe moresfare as 'the use or misuse of values and norms as a weapon of war or as a way to achieve a political objective' (Verweij, 2019a, b, p. 1).

One might argue that the introduction of the term moresfare has little added value, since the term lawfare already covers the most important aspects of the phenomenon in question. However, that is not the case. Notably, statements like these are based on the persistent premise that law, as a tangible body of rules, makes ethics superfluous. Apart from this being a rather dangerous presupposition, it is also not adequate. As I have pointed out elsewhere (Verweij, 2007), ethics is the alpha and omega of law, which means that ethics as a critical reflection on morality (values and norms) precedes law in the sense that the moral deliberation of the members of a community leads to consensus about norms (rules) with regard to behaviour (that is, ethics as the alpha of law). Law and more specifically rules of law are thus crystalized morality. However, rules as crystalized morality require critical reflection (thus ethics) precisely because of their crystalized and therefore petrified and immobile character in an ever-changing context (this is ethics as the omega of law). Critical reflection is needed because in a world characterized by multiplicity and change, the petrification of norms will eventually lead to unacceptable and unwarranted behaviour. The abolition of legislation on discrimination and segregation in South Africa when Apartheid ended in 1990 or the abolition of Jim Crow laws in the south of the US in 1965 is a case in point. A similar case in point, leading however to the opposite effect, were the Nuremberg laws of 1935 in Nazi Germany that legalized the discrimination of Jews. According to Arendt, this phenomenon could occur because 'everybody was sound asleep' when fundamental human values were replaced by their opposites (Arendt, 2012, p. 227). What these concrete examples of legislation on discrimination demonstrate, apart from their completely opposite intentions and moral implications (or lack of attention for these intentions and implications, as Arendt rightly points out), is that legislation changes because morality changes. And morality changes because of the moral deliberation on the values and norms that the moral agents of a specific community hold and that make a community an actual community. There is no community without the shared values and norms that hold this community together in a performative design of these values and norms. In other words, the values and norms that emerge in behaviour and language constitute the community and form the basis of its legislation. Thus, law is inherently connected to ethics; there is no law without ethics (without a moral basis, however small that may be). There are, however, ethics without law in the sense of morality, or moral deliberation, that has not yet turned into law or will not turn into law. It is precisely this moral deliberation between moral agents that can influence, change, strengthen and weaken legislation and can in fact create it or even kill it, as Arendt's example of the replacement of values by the Nazi regime shows.

The added value of the term moresfare is therefore that it specifically addresses this dynamic interaction between ethics and law in the context of hybrid warfare and thereby helps us to better understand hybrid warfare and its consequences. Given the fact that some people believe that law is a tangible body of rules that makes ethics superfluous, it may not come as a surprise that this dynamic between ethics and law is often overlooked. This is also the case in discussions on hybrid warfare, and especially in discussions on lawfare as a form of hybrid warfare. As relevant as these discussions may be, they lead to a one-sided and therefore too limited view and interpretation of both lawfare and hybrid warfare. In order to explain my point, I will focus on three texts on lawfare, two written by the author who introduced the concept (Dunlap) and the third written by two authors (Bachmann & Munoz Mosquera, 2018), who discuss the way hybrid warfare challenges the international legal order and who introduce the concepts of a 'Zeusian' and 'Hadesian' form of lawfare.

The Danger of Moral Myopia in Lawfare

It is interesting to note that both Hoffman's definition and description of hybrid warfare and Dunlap's definition and description of lawfare can be characterized as a form of 'enemy thinking', which is an effective form of moresfare when the term 'enemy' is used to create and stimulate fear and insecurity. As explained in section "Hybrid Warfare", moresfare implies that values and emotions connected to these values are used, misused, influenced and manipulated. In this case, 'enmity' and the subsequent actions of the constructed 'enemy' as the extreme other imply a threat to the values held by the community (freedom, security, wellbeing etc.), which leads to emotions such as anxiety, fear and anger, as well as to a willingness to resist and fight back in the community 'under attack'.

The texts of Hoffman and Dunlap consistently refer to 'them' (the enemy with capital letters) against 'us' (law-abiding victims). In both his texts on lawfare (2001, 2009), Dunlap focuses on lawfare used against the US. He refers to 'foes of the US' (2001, p. 4) that 'exploit the vulnerabilities of the US as an open democratic society'. Furthermore, he states that there is 'disturbing evidence that the rule of law is being hijacked' into another way of fighting (lawfare) (2001, p. 2). He mentions 'America's enemies' (2001, p. 5) who 'callously capitalize on collateral damage incidents' (2001, p. 5). To illustrate his point, he refers to heavy weaponry that was sheltered in mosques and the Taliban who placed a tank and anti-aircraft guns under the trees in front of the office of CARE International (2001, p. 5). In his text written in 2009, he mentions the detainee abuse scandal at Abu Ghraib prison in Iraq, which he calls 'the most serious setback since 9/11', and maintains that 'the exploitation of the incident by adversaries allowed it to become the perfect effects-based asymmetrical operation that continues to present difficulties for American forces' (2009, p. 34).

Of course, it is true that hybrid warfare implies this form of lawfare, meaning that professional soldiers are challenged to break their own rules. Yet, what Dunlap fails to see is the moral dimension that bears heavily on these situations. He fails to see his own use of moresfare, as discussed above, and the moresfare underlying the lawfare that he accuses 'the enemy' of. This last point becomes clear in another example he uses, namely the occupation of a bridge by voluntary human shields that occurred during the war in the Balkans (op. cit., p. 9). By placing themselves on the bridge, people tried to prevent it from being bombed. Dunlap calls this situation 'politically complex, but not legally difficult' (op. cit., p. 9). Dunlap has no eye for the ethical dimension of the situation. The situation is not just about loss of face for politicians, it is about the values held by a community and maybe also by the soldiers who are ordered to bomb the bridge, values such as legitimacy, humanity and justice. The situation is morally/ethically complex and is subsequently politically complex because people are aware of the values that are at stake and are in danger of being violated. What is clear is that Dunlap misses or overlooks the ethical dimension of the situation and thus also misses the effects thereof.

This also holds for the other examples, such as Abu Ghraib. As mentioned above, Dunlap calls the detainee abuse scandal at Abu Ghraib prison in Iraq the most serious setback since 9/11 and maintains that 'the exploitation of the incident by adversaries allowed it to become the perfect effects-based asymmetrical operation that continues to present difficulties for American forces' (p. 34). That may be true, but an incident can only be exploited when the values that people attach great importance to are violated. If there were no violation of values in the first place, there would be no incident that could be exploited.

A similar moral myopia is present in the example of air force lawyers who were criticized for questioning the proposal to strike a statue of Saddam Hussein. Dunlap states that 'no lawyer ever concluded that bombing the statue was illegal; they only recommended that the target was carefully screened for conformance with existing legal standards' (op. cit., p. 23). Yet, what neither Dunlap nor the lawyers in question seem to acknowledge is the ethical dimension of the situation and thus the value of the statue in the eyes of the Iraqi people. As General Horner aptly stated with regard to the bombing of the Baath Party Headquarters, 'In the final analysis we looked at it through American eyes, which was wrong' (op. cit., p. 9). This moral myopia is precisely the point: not seeing all values at stake, especially not the values of the other stakeholders. The same holds for the 'manipulation' of values. Dunlap quotes Reisman and Antoniou, who state that 'public support can erode or even reverse itself rapidly, no matter how worthy the political objective, if people believe that the war is being conducted in an unfair, inhumane, or iniquitous way' (p. 35). Notably, people have these beliefs when the values that are at stake (in this case justice and humanity) are actually being violated. It is obvious that there is hardly any manipulation necessary when people are already convinced of the injustice of specific acts, or maybe even the war as such. People are moral agents. In that capacity, they by definition have moral judgements. And in a democratic society they obviously also have moral judgements about the political leaders that they elected to serve the community that they belong to and the military leaders that they, as a community, entrusted with the monopoly of violence.

These are important things to keep in mind when lawfare (and thus the underlying moresfare) is looked upon as something that can also be used in a 'good' way. Dunlap states that lawfare is not only something that 'adversaries seek to use'. '[A]s the new counterinsurgency doctrine emphasizes', it is also 'a resource that democratic militaries can and should employ affirmatively' (p. 35). He mentions, for instance, legal weaponry to attack the financial networks of terrorist organizations and other sanctions (p. 35). This double use of lawfare—perceived as done in a positive and constructive way or in a negative and destructive way—is also discussed by Bachmann and Munoz Mosquera (2018). They argue that lawfare is ideally suited to be used as part of successful defensive or offensive comprehensive approaches to twenty-first century conflicts (p. 62) and that hybrid warfare offers a platform for the employment of different means of power (pp. 65–66). The use of law as a weapon is one of them, and like other weapons, the law can be used maliciously or affirmatively (p. 65). Bachmann and Munoz Mosquera maintain that there is a so-called proactive, positive 'Zeusian' use of lawfare aimed at neutralizing the malicious 'Hadesian' use of the law (p. 41).

However, this is more complex than it may seem. Again, this has everything to do with the values that are involved in lawfare. Misusing these values for political and strategic purposes is precisely what is meant by moresfare (see section "Moresfare and the Dynamic Between Ethics and Law"). This becomes clear in the example both authors give with regard to the misuse of the term 'peacekeeping', 'humanitarian catastrophe' and 'genocide' as a pretext to invade a country, whilst the underlying and hidden motives for the invasion are political and strategic gain (p. 69). By using these terms (and thus misguiding people), military deployment is justified and given the appearance of legitimacy. However, there is no legitimacy from a philosophical just war perspective. Although the just war tradition is not mentioned by Bachmann and Munoz Mosquera, just war principles might also be relevant in the context of hybrid warfare, given the just war tradition's aim of safeguarding peace, security and humanity (regarding the aims and relevance of the just war tradition, see, for instance, Eberle, 2016; Johnson, 2013; Lazar, 2017; Reichberg et al., 2006; Walzer, 1977). The misuse of these terms and their underlying values means putting aside the most important conditions for a just war ('just cause', 'right intent' and 'last resort').

The 'just cause' principle implies that a war can only be called just if it is waged to defend against aggression that is disturbing peace and security. The 'justness' of a cause is closely connected to the 'right intent' principle, meaning that there are no hidden motives (such as economic, strategic or political gain) and that the cause for which a war is waged is actually 'just' in the sense that the aim is to restore the peace and security that were destroyed or maintain the peace and security that were at risk of being destroyed. The third principle mentioned above, which also relates to the 'justness' of a cause, implies that before actually deciding to go to war, everything must have been done to avoid the destructiveness of a war (such as sanctions, diplomacy etc.). This is precisely what the principle of 'last resort' makes clear. As mentioned above, the philosophical just war tradition, to which these three principles—along with some other principles—belong, was established to actually safeguard values such as humanity and peace and security (see also Verweij, 2019a, b).

When considering the defensive or Zeusian affirmative use of lawfare, as Bachmann and Munoz Mosquera call it, it is important to take the original philosophical just war criteria as a guideline. This is because, as discussed above, the moral/ethical foundation of the just war criteria is the best guarantee for justice and legitimacy. Notably, legitimacy is not the same as legality. Moreover, there is a crucial difference between legality and legitimacy that is directly related to the difference between ethics and law as discussed in section "Moresfare and the Dynamic Between Ethics and Law" (ethics as the alpha and omega of law). A legal order can and should obviously be legal, which means that it is in accordance with the law. However, this does not automatically imply that it is legitimate and thus in accordance with the underlying principles of justice (see also Cliteur & Ellian, 2016). Thus, focusing solely on legality, as Hoffman and Dunlap and to a lesser extent Bachmann and Munoz Mosquera do, means missing the broader moral/ethical point of view. This moral myopia leads to what Honneth calls 'reification' ('Verdinglichung' in German) (Honneth, 2008). Reification, or losing sight of the humanity of the other, undermines the mutual recognition on which respect is based, as Honneth convincingly demonstrates. This is exactly what is at stake in the discussions of lawfare referred to above.

When, as stated by the authors discussed in this section, a community should not only be protected against the lawfare used by 'adversaries' but should also use lawfare as a 'new counterinsurgency doctrine', it is crucial to acknowledge the ethical dimension and thus have a keen eye for the values that are at stake or will be at stake in the situation one either wants to realize or wants to avoid. Insight into this ethical dimension also implies insight into lawfare and the moresfare that it entails. Both lawfare as 'the strategy of using or misusing law as a substitute for traditional military means to achieve an operational objective' (Dunlap, see section "The Danger of Moral Myopia in Lawfare") and moresfare as 'the use or misuse of values and norms as a weapon of war or as a way to achieve a political objective' (Verweij, see section "The Danger of Moral Myopia in Lawfare") are questionable hybrid strategies. They need guidelines and criteria (like the philosophical just war criteria discussed in this section) to keep them from derailing and affecting intrinsic values.

This claim regarding the necessity of keeping a keen eye on the systemic ethical dimension of the hybrid use of force is further substantiated in the next section by a discussion of the concept of resilience, for legitimacy (being in accordance with the underlying principles of justice) is not only a sine qua non for lawfare and moresfare, but, as Pounds et al. (2018) maintain, also for state resilience.

Countering Hybrid Threats: Resilience as a Panacea?

As indicated in the introduction, strengthening the resilience of a community is seen as crucial in countering hybrid warfare and thus the lawfare and moresfare that it entails. It is interesting to note that the concept of resilience has become a buzzword, and the literature on this concept is enormous. This section does not aim to review all of this literature and the different definitions used. Not only has this been done by many others (see, for instance, Atkinson et al., 2009; Kaplan, 1999; Zolli & Healy, 2013) but it would also miss the intended purpose of this chapter. As explained below, resilience can best be described as 'the capacity of a system, enterprise, or a person to maintain its core purpose and integrity in the face of dramatically changed circumstances' (Zolli & Healy, 2013, p. 7). Transferred to a sociopolitical context, resilience is seen as the strength of a community and its members and, in that capacity, it is considered by both NATO and the EU to counteract the hybrid use of force and the effects of lawfare. As discussed below, against this sociopolitical background, the resilience of a community or society implies, amongst others things, 'a high level of the competence of a society in critical thinking and the understanding of the nature of hybrid war instruments'; 'the trust of a society in the integrity of the political system'; 'a strong sense of belonging to a community, citizen empowerment and economic equity', which will reduce the potential for division and the polarization of a society and the countering of various groups of society against each other and against the nation's institutions (Filip, op. cit., p. 67).

It is obvious that this kind of resilience is not automatic. This section aims to show that stimulating and maintaining this form of community resilience implies insight into the systemic ethical dimension. When this insight is absent, there is no clear view on the scope and pitfalls of the concept resilience, let alone on the scope and pitfalls regarding the implementation of this concept. When this is the case, the challenges of hybrid warfare will not be countered. Rather, they will increase, which is exactly what happens when the term resilience is used in a thoughtless and uncritical way. As discussed in the last part of this section, a blind spot is created when it comes to the paradox of resilience.

Resilience as a Personal and Systemic Ideal

It is interesting to note that in a military context resilience is often seen as a disposition that a soldier should aspire to, for instance by developing a stoic attitude (see, for instance, Sherman, 2005). Yet, as the EU and NATO make clear, whole communities need to become resilient in our day and age.

The term resilience is derived from the Latin 'resilire', which refers to the ability to bounce back. However, what does 'bouncing back' imply? As indicated above, there is an abundant use of the term resilience in many disciplines. It is not only used in psychology but also in education, healthcare, the public sector and many other contexts (see, for instance, Atkinson et al., 2009; Kaplan, 1999; Zolli & Healy, 2013). Obviously, this complicates a precise definition of the term. Yet, as different as the descriptions of resilience may seem, they have some essential points in common: continuity, recovery and learning from and in a changing situation. Zolli and Healy, who discuss the concept of resilience in both systems and people, describe resilience in ecological as well as sociological terms as 'the capacity of a system, enterprise, or a person to maintain its core purpose and integrity in the face of dramatically changed circumstances' (Zolli & Healy, p. 7). This definition corresponds to that of Atkinson et al. (2009, p. 137), who define resilience as 'the ability to apparently recover from the extremes of trauma, deprivation, threat or stress'.

As a psychological process, resilience concerns reconstructing and re-modelling the self, which implies learning and changing. Malabou (2011), who discusses the concept of resilience as recuperative power in the work of Cyrulnik, points to the high degree to which learning and change are in fact possible. Cyrulnik studied several cases of Romanian orphans who managed to constructively survive the traumatizing institutions of the Ceausescu period. With the help (care) provided by their foster families, they managed to overcome the severe impact of their tragic experiences. Malabou discusses healing the 'tracks' in their brain, the 'markers' of premature affective neglect and the contempt of society. This ties in with Zolli and Healy's comment that recent scientific research suggests that personal, psychological resilience is more widespread, improvable and teachable than previously thought. The reason for this is that resilience is not only 'rooted in our character, experiences, values, and genes, but critically in our habits of mind—habits we can cultivate and change' (Zolli & Healy, 2013, p. 14).

As these descriptions indicate, resilience is often seen as a positive disposition. In the words of Meadows (2008, p. 76), who discusses resilience on a broader, system level, 'Resilience is a measure of a system's ability to survive and persist within a variable environment [...] it arises from a rich structure of many feedback loops that can work in different ways to restore a system even after a large perturbation'.

Against the background of these positive descriptions, it is no surprise that both the EU and NATO have pinned their hopes on resilience when it comes to countering hybrid warfare attacks. During the NATO Warsaw Summit in 2016, cyberspace was added to the traditional areas of operational warfare and, with it, a commitment to 'increase resilience'. This 'requires that all the Alliance's members maintain and further develop their individual capacity and collective efforts to withstand any form of armed attack. In this context, we are today committed to continuing to improve our resilience to the full range of threats, including hybrid threats, from any direction for a credible defense and effective fulfilment of the Alliance's main tasks' (Filip, 2017).

According to Filip (2017), NATO's focus on resilience is based on two trends: the increasing dependence of today's armed forces on the capabilities and infrastructures that are owned or operated by the civilian sector and, secondly, the vulnerability of this same sector to external attacks or domestic problems that can be exploited by 'potential opponents' (op. cit., p. 65). However, this exploitation cannot take place when the adversary in question is dealing with a resilient society. In the words of Filip, 'Hybrid war strategy—is essentially a strategy designed to cause disturbances, confusion, destabilization, and paralysis (for example, modelling the behavior of the target nation)—and can be countered by demonstrating that all of these goals are not achieved due to the strength of the target' (op. cit., p. 67).

What does this strength (resilience) imply? Filip mentions, amongst others things, 'a high level of the competence of a society in critical thinking and the understanding of the nature of hybrid war instruments'; 'the trust of a society in the integrity of the political system'; 'a strong sense of belonging to a community, citizen empowerment and economic equity', which will reduce the 'potential for division and polarization of a society and the countering of various groups of society against the others and against the nation's institutions' (op. cit., p. 67). In a similar way, the EU states the following in its Global Strategy: 'It is in the interests of our citizens to invest in the resilience of states and societies to the east stretching into Central Asia, and south down to Central Africa. A resilient society featuring democracy, trust in institutions, and sustainable development lies at the heart of a resilient state' (Shared Vision, Common Action: A Stronger Europe, A Global Strategy for the European Union's Foreign and Security Policy).

It is thus clear that a resilient society presupposes quite a bit. It can only arise when certain conditions are met. In fact, one can hear the echo of Freud's emphasis on the necessity of emotional bonds (see section "Violence, Power and Law and the Role of Emotional Bonds"), for what is basically needed is trust in the political system in question and a sense of belonging to society as a community (see also Zolli and Healy [2013] with regard to the critical role of trust and cooperation). Moreover, as was pointed out by Filip, this implies citizen empowerment (including the ability to think critically) and economic equity. So again, when political leaders wish to have a resilient society, an old but highly relevant adage should be heeded, namely the *tranquillitas ordinis*, as Augustine, the 'father' of the just war tradition, called it:

'peace or tranquility ensured by good governance in the political realm' (Augustine cited by Reichberg et al., 2006, p. 177). In other words, a society can only be resilient when the people who make a society a community experience it as just and thus well governed. It hardly needs saying that this is not automatic and requires considerable effort in our current Zeitgeist, which brings me to the paradox of resilience.

The Paradox of Resilience

As the previous section shows, the concept of resilience is more complex than it appears to be at first sight. Moreover, an uncritical and thus unreflective use of the term might lead to unwanted and harmful consequences, for resilience might lead to its opposed effect, or might be seen as interchangeable with its counterpart, as several authors have demonstrated. Zolli and Healy (2013), for instance, point out that resilience should not be equated with robustness (op. cit., p. 13) and Meadows (2008, p. 76) maintains that the opposites of resilience are brittleness and rigidity. Similarly, Ogunbode et al. (2019) state that resilience does not imply an absence of vulnerability, and distress and disorder are normal responses to adversity and trauma and do not indicate a lack of resilience (op. cit., p. 705).

Resilience thus implies a certain amount of sensitivity and should not be put on a par with hardness or invulnerability, for this would paradoxically lead to the opposite of resilience and thus to the opposite of what one aspires to. In addition, resilience is a relative concept in the sense that it is related to different contexts and thus to different forms of vulnerability and invulnerability in these contexts. Zolli and Healy (2013, p. 17) rightly point out that resilience is not a virtue in itself, for 'terrorists and criminal organizations are also highly resilient'. A similar point is made by Atkinson et al. (2009) in their discussion of two groups of resilient nurses in the context of the Second World War. Members of both groups put their nursing skills into practice. However, the first group worked in the Auschwitz-Birkenau concentration camp, helping and caring for those in need. The other group of nurses assisted in murdering the patients in their care as part of the 'euthanasia' programme for people with physical or mental disabilities. According to Atkinson et al. (2009), both groups can be considered resilient, yet in different and opposite ways. Resilience can be attributed to members of the first group on the basis of their care for and devotion to people in need despite the inhumane working conditions and to members of the second group on the basis of their dutifulness, planning abilities and focus on survival. The care for others, the compassion and empathy (first group), stands in sharp contrast to the absence thereof (second group).

However, since both groups can be considered resilient, and resilience thus seems to refer to two extremes on a continuum, the first form might paradoxically transform into its opposite. In other words, given specific circumstances and influences, the compassionate and emphatic attitude might paradoxically turn into its opposite and thus into ego-focused hardness, at almost all costs. The paradox of resilience can also be detected in the context of climate change. Ogunbode et al. (2019) discovered that the reduction of emotional responses and the capacity to be able to deal with extreme weather conditions (such as floods) might lead to a diminished motivation for and engagement with actions to counter climate change. The ability to accommodate the problem (the resilient ability) seems to stand in the way of solving the problem (or taking action that contributes to solving the problem).

This paradox is also underlined by Diprose (2014), who argues that a focus on resilience might encourage people to accept a life of insecurity and social struggle and can undermine collective action for change. It places the responsibility for inequality, insecurity and detrimental social conditions with the individual and not with the government.

Resilience can thus transform into its own opposite meaning. In certain situations, it can become the opposite of empowerment or the opposite of empathy and compassion, as was the case with the second group of nurses in the example discussed above. Thus, the resilience that people are praised for might at the same time imply an acceptance of detrimental and destructive circumstances and subsequently of pain and inhumanity and other problematic and inhumane consequences, without efforts being made to change these circumstances. It can disenfranchise people and block their ability and willingness for constructive change. Given this paradox of resilience and the negative moral/ethical consequences it may generate, resilience is clearly not the panacea that it is sometimes considered to be. However, this does not mean that the concept is of no use. As was concluded in section "The Danger of Moral Myopia in Lawfare", both lawfare and moresfare are questionable hybrid strategies. They need guidelines and criteria to keep them from derailing and affecting intrinsic values; they therefore need a systemic ethical approach. This also holds for resilience as a way to counter hybrid strategies. Only resilience built on this systemic approach engenders the trust, cooperation, compassion and respect that is aspired for by the EU and NATO.

Conclusion

In discussions on hybrid warfare and the countering thereof, the connection between violence, power and law and the crucial role of the psychological and ethical dimension are often overlooked, as section "Violence, Power and Law and the Role of Emotional Bonds" made clear. Emotions and the values associated with these emotions strengthen the bonds between the communities at war, or in conflict, either for better or for worse, yet this crucial role of ethics and psychology on a political level is often put aside, minimized or acknowledged only in a limited or unilateral way, as was illustrated by the different assessments of present-day conflicts by Kaldor and Hoffman in section "Hybrid Warfare". This difference between the two authors is caused by the fact that Kaldor includes a broader ethicopolitical approach in her analysis and thus addresses the values that are at stake in the (global) system as a whole, whereas Hoffman primarily focuses on the US (and Western societies) as a victim of aggression. Section "Moresfare and the Dynamic Between Ethics and Law" made clear that this broader ethicopolitical approach helps us to better understand hybrid warfare and specifically the way in which, in hybrid warfare, values and the emotions associated with these values are played upon, influenced, questioned and moulded. This way of using and misusing values and norms was called 'moresfare' and is connected to the more familiar lawfare, as was explained in section "Moresfare and the Dynamic Between Ethics and Law". The concept of 'moresfare' not only contributes to a sound understanding of hybrid warfare but also helps to reveal the interaction between ethics and law, especially in hybrid conflicts, as section "The Danger of Moral Myopia in Lawfare" made clear.

The dynamic between ethics and law is often overlooked, which is caused by the fact that legitimacy and legality are put on a par. Legitimacy is not the same as legality, however. Moreover, placing legality above legitimacy means missing the moral/ethical point of view and therefore also the moresfare beneath lawfare. The analysis of the texts of several authors on hybrid warfare and its lawfare strategies in section "The Danger of Moral Myopia in Lawfare" revealed the moresfare underneath the lawfare and identified them both as questionable strategies that need guidelines and criteria to forestall both strategies from derailing and affecting intrinsic values. The philosophical just war criteria might still be useful in this regard, as was suggested in section "The Danger of Moral Myopia in Lawfare". This means that insight into the systemic ethical dimension is indispensable, and the absence thereof, or moral myopia, can have destructive consequences. This also holds for the countering of hybrid warfare, as section "Countering Hybrid Threats: Resilience as a Panacea?" made clear. Moral myopia implies a blindness to the system as a whole and leads to a thoughtless and uncritical use of the term resilience and the absence of a clear view on the paradox of resilience, as was discussed in section "Countering Hybrid Threats: Resilience as a Panacea?".

The problems this generates should be taken seriously, for both the EU and NATO have pinned their hopes on resilience as a way to counter hybrid warfare. However, the resilient society they aspire to presupposes the emotional bonds Freud addressed and insight into the political dimension of emotions (as section "Violence, Power and Law and the Role of Emotional Bonds" made clear). It also presupposes trust in the political system and its leaders, citizen empowerment (including the ability to think critically) and economic equity, as was discussed in section "Countering Hybrid Threats: Resilience as a Panacea?". Added to that, it calls for attention to be paid to the paradox of resilience, for resilience can develop in its own opposite meaning. The final part of section "Countering Hybrid Threats: Resilience as a Panacea?" showed that it can become the opposite of empowerment and can disenfranchise people.

In order to forestall this, a systemic ethical approach is needed, since that can guarantee that the values and associated interests of all stakeholders and the effects of actions taken, including the long-term effects, are taken into account. This approach is a way of system-thinking that seems indispensable, for resilience built on this systemic approach can engender the trust, cooperation, compassion and respect discussed in section "Countering Hybrid Threats: Resilience as a Panacea?". If this systemic ethical approach is absent, countering hybrid warfare and, in line with that, strengthening the resilience of a community are doomed to fail.

However, resilience does not refer to a fixed point that can be reached once and for all. Rather, it seems to be a moment of balance in a continuum stretching from extreme vulnerability and fragility to invulnerability and hardness; from hypersensitivity to insensitivity. Yet, this whole continuum is part of our human condition. Therefore, the best that we can aspire for in our often messy ethicopolitical endeavours are moments of balance—possibly a shaky balance—between the extremes of the continuum. By definition, this balance contains elements of both sides.

References

Alford, F. (1998). Freud and violence. In A. Elliott (Ed.), *Freud 2000*. Routledge.

Arendt, H. (2012). *Denken. Het Leven van de Geest*. Klement Pelckmans. (Originally: *'Thinking' from The Life of the Mind*, 1971.)

Atkinson, P., Martin, C., & Rankin, J. (2009). Resilience revisited. *Journal of Psychiatric and Mental Health Nursing, 16*(2), 137–145.

Bachmann, S. D., & Munoz Mosquera, A. B. (2018). Hybrid warfare as lawfare: Towards a comprehensive approach. In E. Cusumano & M. Corbe (Eds.), *A civil-military response to hybrid threats*. Palgrave Macmillan.

Cliteur, P., & Ellian, A. (2016). *Legaliteit en legitimiteit: De grondslagen van het recht*. Leiden University Press.

Diprose, K. (2014). Resilience is futile. *Soundings, 58*, 44–56. https://doi.org/10.3898/136266215814379736

Dunlap, C. J. (2001). *Law and military interventions: Preserving humanitarian values in the 21st century conflicts* (Working Paper). Harvard/Boston. www.ksg.harvard.edu/cchrp/Web%20Working%20Papers/Use%20of%20Force/Dunlap2001.pdf

Dunlap, C. J. (2009). Lawfare: A decisive element of 21st century conflicts? *Joint Force Quarterly, 54*, 34–40.

Eberle, C. J. (2016). *Justice and the just war Tradition: human worth, moral formation, and armed conflict*. Routledge.

Ebi, K. L. (2011). Resilience to the health risks of extreme weather events in a changing climate in the United States. *International Journal of Environmental Research and Public Health, 8*(12), 4582–4595. https://doi.org/10.3390/ijerph8124582

Einstein, A., & Freud, S. (2018). *Waarom Oorlog?* van Klaveren. (Original: *Warum Krieg?* 1933.)

Filip, L. (2017). NATO resilience strategy towards Russian hybrid warfare. *Journal of Defense Resources Management, 9*(2), 64–72.

Hoffman, F. (2007). *Conflict in the 21st century: The rise of hybrid wars*. Potomac Institute for Policy Studies.

Honneth, A. (2008). *Reification: A new look at an old idea* (M. Jay, Ed. with contributions from J. Butler, R. Geuss & J. Lear [part of the Berkeley Tanner Lectures]). University Press.

Johnson, J. T. (2013). Contemporary just war thinking: Which is worse, to have friends or critics? *Ethics & International Affairs, 27*(1), 25–45.

Kaldor, M. (2007). *Human security: Reflections on globalization and intervention*. Polity Press.

Kaldor, M. (2008). From just war to just peace. In M. den Boer & J. de Wilde (Eds.), *The viability of human security*. Amsterdam University Press.

Kaplan, H. B. (1999). Toward an understanding of resilience: A critical review of definitions and models. In M. D. Glantz & J. L. Johnson (Eds.), *Resilience and development: Positive life adaptations* (pp. 17–83). Kluwer Academic Publishers.

Lazar, S. (2017). Just war theory: Revisionists versus traditionalists. *Annual Review of Political Science, 20*, 37–54.

Malabou, C. (2011). *Wat te Doen met ons Brein*? Boom. (Originally: *Que Faire de Notre Cerveau*?)
McCulloh, T., & Johnson, R. (2013). *Hybrid warfare*. Joint Special Operations University (JSOU) Report 13-4.
Meadows, D. H. (2008). *Thinking in systems*. Chelsea Green Publishing.
Moisi, D. (2010). *The geopolitics of emotion: How cultures of fear, humiliation, and hope are reshaping the world*. Random House.
Ogunbode, C. A., Böhm, G., Capstick, S. B., Demski, C., Spence, A., & Tausch, N. (2019). The resilience paradox: Flooding experience, coping and climate change mitigation intentions. *Climate Policy, 19*(6), 703–715. https://doi.org/10.1080/14693062.2018.1560242
Pounds, N., Alam, I. E., & Keijzer, R. (2018). A military contribution to state resilience: Afghanistan and Lebanon. In E. Cusumano & M. Corbe (Eds.), *A civil-military response to hybrid threats*. Palgrave Macmillan.
Reichberg, G. M., Syse, H., & Begby, E. (2006). *The ethics of war: Classical and contemporary readings*. Blackwell.
Sherman, N. (2005). *Stoic Warriors: The ancient philosophy behind the military mind*. Oxford University Press.
Stoltenberg, J. (2015). Zero-sum? Russia, Power Politics, and the Post-Cold War Era. In *Brussels Forum, March, 20*. http://www.nato.int/cps/en/natohq/opinions_118347.htm
Verweij, D. E. M. (2007). Ethiek als de alfa en omega van recht: Het belang van ethiek en ethiekonderwijs. In W. M. J. Bekkers, et al. (Eds.), *De Advocatenstandaard*. SDU Uitgevers.
Verweij, D. E. M. (2019a, September 23–24). *Ethics and hybrid warfare: Focusing on the 'moresfare' beneath the 'lawfare'*. Paper presented at the International Workshop 'Hybrid warfare—A challenge for security and peace, legality and ethics', jointly organized by the Institute for Peace Research and Security Policy (IFSH) at the University of Hamburg, the Institute for Theology and Peace (ITHF) and the European Centre of Excellence for Countering Hybrid Threats (Hybrid CoE).
Verweij, D. E. M. (2019b). Over 'caritas' en de belofte van de 'juiste intentie': Terug naar de wortels van rechtvaardigheid in oorlog. In het themanummer Oorlog en Rechtvaardigheid. *Algemeen Nederlands Tijdschrift voor Wijsbegeerte, 111*(1).
Walzer, M. (1977). *Just and unjust wars: A moral argument with historical illustrations*. HarperCollins.
Zolli, A., & Healy, A. M. (2013). *Resilience: Why things bounce back*. Simon & Schuster.

Prof. Désirée Verweij held a chair in Military Ethics at the Netherlands Defence Academy and a chair in Normative and Policy Dilemmas of Multilateral Peace Operations at the Centre for International Conflict—Analysis & Management, Radboud University. Her research includes both fundamental and applied ethics, in particular ethics of military practice at the individual, organizational and political level. The contributions in this book celebrate her achievements and aim to contribute to continuing attention for these important themes.

Concluding Reflections

Tine Molendijk and Eric-Hans Kramer

Inspired by Désirée Verweij's work, the contributions in this book reflected on violence in extreme conditions and on related ethical challenges at the micro, meso and macro levels. The confrontation with violence, it became clear, is not easily captured in language. Quite literally, violence often leaves us speechless. Theoretically, too, violence may be defined as the opposite of language, rationality and meaningfulness. At the same time, language, rationality and meaningfulness are conjoined with violence (chapter "Language and Violence"). That is, although we may find that no adequate words exist for the visceral confrontation with violence, we do speak about it, which, then again, we only seem to be able to do in ways that obscure its bloody reality or that attribute it to the Other. For the same reasons that we are repelled by violence, we are also fascinated by it. We rationally use violence to end violence, or sometimes use it just for the sake of it, and technical rationality may at once create and rationalize brutal forms of violence. Violence, in a way, is always meaningful. The specific ways in which we perpetrate, witness or suffer violence are highly significant to us, hence our efforts to use and speak about it in ways that give us justification, or conversely to hurt our victims in such ways that it augments their suffering.

Put differently, whilst violence might be defined as essentially the opposite of what defines humanity, as 'inhumane', at the very same time only human beings are able to be 'inhumane' in this way (chapters "Language and Violence" and "Military Trauma and the Conflicted Human Condition: Moral Injury as a Window into Violence, Human Nature and Military Ethics"). This also complicates the notion of morality as the humane, rationalist opposite of the bestial irrationality of violence. Violence is not something outside of us as humans but within us, and humans are not

T. Molendijk (✉) · E.-H. Kramer
Department of Military Management Studies, Faculty of Military Sciences, Netherlands Defense Academy, Breda, The Netherlands

E.-H. Kramer
e-mail: fj.kramer@mindef.nl; erichans@xs4all.nl

E.-H. Kramer and T. Molendijk (eds.), *Violence in Extreme Conditions*,
https://doi.org/10.1007/978-3-031-16119-3_13

just rational actors but meaning-making, emotion-experiencing beings. Accordingly, moral intuitions and emotions play a crucial role in moral judgement (chapters "The E-Word (Emotions) in Military Moral Education: Making Use of the Dual-Process Model of Moral Psychology" and "'Moresfare' and the Resilience Paradox: Ethics as the Terra Incognita of Hybrid Warfare and Its Challenges"). In fact, rational arguments may even be used instrumentally to justify the use of force (chapter "Instrumental Morality Under a Gaze: Israeli Soldiers' Reasonings on Doing "Good""). More generally, humans tend to use all kinds of unconscious psychological and social defences against the anxiety of the potential confrontation with violence, with varying degrees of success (chapter "Exploring the Relevance of the Systems Psychodynamic Approach to Military Organizations"). Note, however, that this view on violence and morality is not necessarily a pessimistic one. There is nothing inherently worrisome about the given that not just rational arguments but also moral emotions such as disgust, shame and guilt drive our behaviour, and the fact that we feel the need to rationalize violence also underlines the fact that we are unable to simply accept the harming of people. We need violence to occur in a way that gives us justification. If we cannot see our own and others' actions as such, if they become unjust, random and meaningless, they become sources of anxiety and moral injury (chapters "Military Trauma and the Conflicted Human Condition: Moral Injury as a Window into Violence, Human Nature and Military Ethics" and "Exploring the Relevance of the Systems Psychodynamic Approach to Military Organizations").

Violence is the distinguishing characteristic, the raison d'être, of the military organization. The last decades have seen significant changes in the nature of military intervention. Conflicts all over the world are characterized by increasingly complex dynamics, whilst the international community faces increasingly demanding interpretations of its moral responsibilities with regard to these conflicts. Accordingly, contemporary military intervention increasingly involves the use of methods such as diplomacy and development (chapter "Does the Comprehensive Approach Have Any Future as a Strategy for Intervention?") and a focus on post-conflict peace through reconstruction and reconciliation (chapter "Contemporary Just War Thinking and Military Education"). Consequently, military personnel are often asked to exercise restraint, even when their opponents do not (chapter "Educating for Restraint"). Furthermore, many armed forces have adopted more decentralized command structures in which subordinate commanders are given both greater freedom of action and greater responsibility (chapter "Soldiers as Street-Level Bureaucrats? Military Discretionary Autonomy and Moral Professionalism in a Police Perspective"). The soldiers of today thus have to decide on when and how to apply coercion or exercise restraint in unclear situations.

How to deal with these developments? Whilst the authors in this book come from different backgrounds and work in different fields, there are remarkable similarities in their answers to this question. They all arrive at answers that are practically oriented whilst maintaining a central role for ethics. In fact, it seems to be exactly because of the authors' concern for the real-world interests of all stakeholders involved in military practice, including military personnel themselves, that they emphasize the importance of persistent commitment to ethical standards. At the same time, they seek *feasible* ways of maintaining that commitment. Furthermore, their aspiration to really grasp the complicated actuality of military practice has led many of them to draw on multiple disciplinary perspectives. Moreover, whilst virtually all authors are attentive of recent developments in military practice, they appear to conclude that the shift towards 'new wars' does not mean that older wisdoms cease to apply. Some demonstrate that more traditional approaches may even be more holistic and interdisciplinary than modern theories.

To be more specific, the contributions in this volume argue that today's world requires an orientation towards a human security approach and related strategies of development and diplomacy (chapter "Does the Comprehensive Approach Have Any Future as a Strategy for Intervention?"), a more comprehensive notion of the post-war peace we aspire to achieve through military intervention (chapter "Educating for Restraint"), more emphasis on restraint and less on warriorship in the moral education of soldiers (chapter "Contemporary Just War Thinking and Military Education"), and more attention for the role that emotion and intuition play in moral judgement (chapter "The E-Word (Emotions) in Military Moral Education: Making Use of the Dual-Process Model of Moral Psychology"), all the while paying attention to not only individual competences but the ethical climate and organizational design as well (chapter "The Dutch Approach to Ethics: Integrity Management in the Military"), and whilst keeping an eye on the misuse of moral arguments as a weapon in war and to conceal other motives (chapters "Instrumental Morality Under a Gaze: Israeli Soldiers' Reasonings on Doing "Good"" and "'Moresfare' and the Resilience Paradox: Ethics as the Terra Incognita of Hybrid Warfare and Its Challenges"). At the same time, the contributions argue that human security should be pragmatically approached as a welcome addition rather than as an alternative to state security strategies (chapter "Does the Comprehensive Approach Have Any Future as a Strategy for Intervention?"), that we should not make just war theory an overly idealist vision but preserve it as useful guidance for real-world problems (chapter "Contemporary Just War Thinking and Military Education"), that the decreasing relevance of martial virtues does not necessarily diminish the relevance of existing ethics theories in themselves (chapter "Contemporary Just War Thinking and Military Education"), that rational reasoning remains a vital element in ethics education as it allows soldiers to work on their moral emotions and intuitions (chapter "The E-Word (Emotions) in Military Moral Education: Making Use of the Dual-Process Model of Moral Psychology"), and that the misuse of moral arguments such as 'moresfare' actually underscores the continuing relevance of a holistic ethical approach to military intervention (chapters "The Dutch Approach to Ethics: Integrity Management in the Military" and "'Moresfare' and the Resilience Paradox: Ethics as the Terra Incognita of Hybrid Warfare and Its Challenges").

To come back to Désirée Verweij's observation with which we opened this book: 'Nobody at the military academy ever talks about violence'. In this volume, we did try to talk about violence, and to explicate that which tends to get hidden in the language of technical rationality. But did we *really* talk about violence? In many cases, we 'merely' discussed violence in terms of what it does as an instrument. In fact, we emphasized that military intervention actually increasingly comprises 'operations other than war' and strategies other than combat, such as development and diplomacy. That said, we also signalled that such activities may unintentionally create new local, oppressive dynamics and may actually imply other forms of Western (moral) imperialism. Moreover, we observed that non-violent discourse can be used as a covert weapon of war. Indeed, many of the so-called non-violent military activities of today can be seen as a continuation of violence by other means, to paraphrase Clausewitz's often quoted dictum that war is a 'continuation of policy by other means' (which, notably, Clausewitz wrote to refine a claim he found to be just as true, namely that war is 'an act of violence intended to compel our opponent to fulfil our will'). Thus, in this volume, we blurred the divide between violence and non-violence.

To state that violence is gradual is not to suggest that just about anything can be interpreted as violence. To be sure, the graduality and the relativity of violence should not be conflated. Rather, we showed that violence manifests itself in more ways than is clear at first sight. Violence can be physical, psychological, structural and symbolic, and it has varying degrees. Naming ostensibly different manifestations as violence is important in order to understand parallels in their dynamics and impact. In fact, doing so can help to protect the concept of violence from an all-encompassing inflation that would make it lose all meaning. Instead, it allows us to better define and demarcate the specific mechanisms of violence.

These considerations on violence also have implications for military ethics. For instance, the shift towards operations other than war implies that military ethics should become less focused on warriorship, whilst the insight that these operations do still involve forms of violence and human suffering suggests that ethics education for soldiers still cannot be the same as ethics training for, say, local government officers. It is important to maintain balance. A bigger, underlying issue is that whilst measures such as ethics training, codes of conduct, integrity committees and 'just war' criteria are supposed to strengthen the moral competence of militaries, they may also become an impossible burden with adverse consequences. In the worst case, soldiers end up in a double-bind 'damned if you do, damned if you don't' position of being blamed for either rule rigidity or incompliance. Military ethics, then, may even become 'moresfare', a strategy to appease society that conceals rather than addresses the morally complex reality of military intervention. Therefore, the changes that militaries have made to address existing ethical issues, under the scrutinizing eyes of society, may both solve old challenges and create new ones, including new responsibilities, dilemmas and transgressions. We have made a start on reflecting on these challenges, from the micro to the macro level, and invite the reader to participate in the discussion.

Dr. Tine Molendijk is an interdisciplinary-minded cultural anthropologist specialized in the topics of violence, military culture and mental health, in particular post-traumatic stress disorder (PTSD) and moral injury. Her Ph.D. research focused on the role of political practices and public perceptions in moral injury in Bosnia and Afghanistan veterans. Currently, she works as Assistant Professor at the Faculty of Military Sciences at the Netherlands Defence Academy (NLDA) and as a Research Fellow affiliated to the Radboud University. Among her publications is the award-winning book *Moral Injury and Soldiers in Conflict: Political Practices and Public Perceptions* (Routledge, 2021).

Prof. Eric-Hans Kramer is Professor in Management and Organization Studies at the Netherlands Defence Academy. Before that, he was associate professor Human Factors and Systems Safety at the same organization. He is involved in several research projects into organizational issues the military organization is being confronted with and with the development of research and educational programs into the psychosocial dynamics in military organizations.

MIX
Papier aus verantwortungsvollen Quellen
Paper from responsible sources
FSC® C105338

If you have any concerns about our products,
you can contact us on
ProductSafety@springernature.com

In case Publisher is established outside the EU,
the EU authorized representative is:
Springer Nature Customer Service Center GmbH
Europaplatz 3, 69115 Heidelberg, Germany

Printed by Libri Plureos GmbH
in Hamburg, Germany